# Security and Privacy Issues in Sensor Networks and IoT

Priyanka Ahlawat
*National Institute of Technology, Kurukshetra, India*

Mayank Dave
*National Institute of Technology, Kurukshetra, India*

A volume in the Advances in
Information Security, Privacy, and
Ethics (AISPE) Book Series

Published in the United States of America by
    IGI Global
    Information Science Reference (an imprint of IGI Global)
    701 E. Chocolate Avenue
    Hershey PA, USA 17033
    Tel: 717-533-8845
    Fax: 717-533-8661
    E-mail: cust@igi-global.com
    Web site: http://www.igi-global.com

Library of Congress Cataloging-in-Publication Data

Names: Ahlawat, Priyanka, 1982- editor. | Dave, Mayank, 1967- editor.
Title: Security and privacy issues in sensor networks and IoT / Priyanka
  Ahlawat and Mayank Dave, editors.
Description: Hershey, PA : Information Science Reference, [2020] | Includes
  bibliographical references. | Summary: "This book examines developments,
  challenges, and research trends in security and privacy issues in
  sensors networks and the internet of things"-- Provided by publisher.
Identifiers: LCCN 2019021550 | ISBN 9781799803737 (h/c) | ISBN
  9781799803751 (eISBN) | ISBN 9781799803744 (s/c)
Subjects: LCSH: Internet of things--Security measures. | Sensor
  networks--Security measures.
Classification: LCC TK5105.8857 .S426 2020 | DDC 005.8--dc23
LC record available at https://lccn.loc.gov/2019021550

This book is published in the IGI Global book series Advances in Information Security, Privacy,
and Ethics (AISPE) (ISSN: 1948-9730; eISSN: 1948-9749)

British Cataloguing in Publication Data
A Cataloguing in Publication record for this book is available from the British Library.

All work contributed to this book is new, previously-unpublished material.
The views expressed in this book are those of the authors, but not necessarily of the publisher.

For electronic access to this publication, please contact: eresources@igi-global.com.

# Advances in Information Security, Privacy, and Ethics (AISPE) Book Series

ISSN:1948-9730
EISSN:1948-9749

Editor-in-Chief: Manish Gupta, State University of New York, USA

## MISSION

As digital technologies become more pervasive in everyday life and the Internet is utilized in ever increasing ways by both private and public entities, concern over digital threats becomes more prevalent.

The **Advances in Information Security, Privacy, & Ethics (AISPE) Book Series** provides cutting-edge research on the protection and misuse of information and technology across various industries and settings. Comprised of scholarly research on topics such as identity management, cryptography, system security, authentication, and data protection, this book series is ideal for reference by IT professionals, academicians, and upper-level students.

## COVERAGE

- Global Privacy Concerns
- Privacy-Enhancing Technologies
- Technoethics
- Cyberethics
- IT Risk
- Cookies
- Internet Governance
- Telecommunications Regulations
- Risk Management
- Device Fingerprinting

IGI Global is currently accepting manuscripts for publication within this series. To submit a proposal for a volume in this series, please contact our Acquisition Editors at Acquisitions@igi-global.com or visit: http://www.igi-global.com/publish/.

# Titles in this Series

*For a list of additional titles in this series, please visit:*
*https://www.igi-global.com/book-series/advances-information-security-privacy-ethics/37157*

### Security, Privacy, and Forensics Issues in Big Data
Ramesh C. Joshi (Graphic Era University, Dehradun, India) and Brij B. Gupta (National Institute of Technology Kurukshetra, India)
Information Science Reference • ©2020 • 456pp • H/C (ISBN: 9781522597421) • US $235.00

### Handbook of Research on Machine and Deep Learning Applications for Cyber Security
Padmavathi Ganapathi (Avinashilingam Institute for Home Science and Higher Education for Women, India) and D. Shanmugapriya (Avinashilingam Institute for Home Science and Higher Education for Women, India)
Information Science Reference • ©2020 • 482pp • H/C (ISBN: 9781522596110) • US $295.00

### Advanced Digital Image Steganography Using LSB, PVD, and EMD Emerging Research and Opportunities
Gandharba Swain (Koneru Lakshmaiah Education Foundation, India)
Information Science Reference • ©2019 • 201pp • H/C (ISBN: 9781522575160) • US $165.00

### Developments in Information Security and Cybernetic Wars
Muhammad Sarfraz (Kuwait University, Kuwait)
Information Science Reference • ©2019 • 351pp • H/C (ISBN: 9781522583042) • US $225.00

### Cybersecurity Education for Awareness and Compliance
Ismini Vasileiou (University of Plymouth, UK) and Steven Furnell (University of Plymouth, UK)
Information Science Reference • ©2019 • 306pp • H/C (ISBN: 9781522578475) • US $195.00

*For an entire list of titles in this series, please visit:*
*https://www.igi-global.com/book-series/advances-information-security-privacy-ethics/37157*

701 East Chocolate Avenue, Hershey, PA 17033, USA
Tel: 717-533-8845 x100 • Fax: 717-533-8661
E-Mail: cust@igi-global.com • www.igi-global.com

# Editorial Advisory Board

# Table of Contents

# Detailed Table of Contents

    *Sachin Kumar Gupta, Shri Mata Vaishno Devi University, India*
    *Aabid Rashid Wani, Shri Mata Vaishno Devi University, India*
    *Santosh Kumar, Department of Computer Science and Engineering, Dr.*
       *SPM IIIT Naya Raipur, Chhattisgarh, India*
    *Ashutosh Srivastava, RST Ecoenergy Private Limited, Mirzapur, India*
    *Diwankshi Sharma, Shri Mata Vaishno Devi University, India*

Due to suppression of central administration in WMN, network functioning like network controls, management, routing, switching, packet forwarding etc. are distributed among nodes, either collectively or individually. So, cooperation among nodes is highly solicited. However, there may exist node's malicious activities because of its open characteristics and limited available battery power. The nodes may misbehave by refusing to provide service or dropping down the packets because of its selfishness and malicious activity. The identification of misbehaving nodes and prevention from them can be one of the biggest challenges. Hence, the prime target of the chapter is to provide an overview of existing intrusion detection and prevention approaches, and secure routing or framework that can recognize and prevent from the malicious activities. The digital signature-based IDS to offer secure acknowledgment and an authentication mechanism has also been discussed. The expectation is the digital signature-based IDS will overcome the weakness of existing IDS.

  *Samyak Jain, National Institute of Technology Karnataka, Surathkal,*
    *India*
  *K. Chandrasekaran, National Institute of Technology Karnataka,*
    *Surathkal, India*

This chapter presents a comprehensive view of Industrial Automation using internet of things (IIoT). Advanced Industries are ushering in a new age of physical production backed by the information-based economy. The term Industrie 4.0 refers to the 4th paradigm shift in production, in which intelligent manufacturing technology is interconnected with physical machines. IIoT is basically a convergence of industrial systems with advanced, near-real-time computing and analytics, powered by low cost and low power sensing devices leveraging global internet connectivity. The key benefits of Industrial IoT systems are a) improved operational efficiency and productivity b) reduced maintenance costs c) improved asset utilization, monitoring and maintenance d) development of new business models e) product innovation and f) enhanced safety. Key parameters that impact Industrial Automation are a) Security b) Data Integrity c) Interoperability d) Latency e) Scalability, Reliability, and Availability f) Fault tolerance and Safety, and g) Maintainability, Serviceability, and Programmability.

  *Kavi Priya S., Mepco Schlnek Engineering College, India*
  *Vignesh Saravanan K., Ramco Institute of Technology, Rajapalayam,*
    *India*
  *Vijayalakshmi K., Ramco Institute of Technology, Rajapalayam, India*

Evolving technologies involve numerous IoT-enabled smart devices that are connected 24-7 to the internet. Existing surveys propose there are 6 billion devices on the internet and it will increase to 20 billion devices within a few years. Energy conservation, capacity, and computational speed plays an essential part in these smart devices, and they are vulnerable to a wide range of security attack challenges. Major concerns still lurk around the IoT ecosystem due to security threats. Major IoT security concerns are Denial of service(DoS), Sensitive Data Exposure, Unauthorized Device Access, etc. The main motivation of this chapter is to brief all the security issues existing in the internet of things (IoT) along with an analysis of the privacy issues. The chapter mainly focuses on the security loopholes arising from the information exchange technologies used in internet of things and discusses IoT security solutions based on machine learning techniques including supervised learning, unsupervised learning, and reinforcement learning.

Body area networks (BANs), a type of Personal Area Networks (PANs), form a significant part of health care applications. This chapter analyzes the effect of channel modeling on the intercept behavior of a wireless BAN while taking optimal sensor scheduling into account. A comparison is drawn between Lognormal and Weibull models for this case. Wireless BANs represent wireless networks of sensors allocated on, in, and around the human body. BANs are basically meant for health care applications where long-lasting and reliable operation is a must. Some healthcare applications carry sensitive information, therefore security is an important issue. A BAN with a sink node and various sensors is considered here along with an eavesdropper. Due to the radio wave propagation's broadcast nature, the wireless communication can be overheard by the eavesdropper. To safeguard the BAN, the propagation channels need to be characterized and modeled for designing reliable communication systems.

# Chapter 5

In wireless sensor network, node localization is helpful in reporting the event's origin, assisting querying of sensors, routing, and various cyber-physical system applications, where sensors are required to report geographically meaningful data for location-based applications. One of the accurate ways of localization is the use of anchor nodes which are generally equipped with global positioning system. However, in range-based approaches used in literature, like Angle of Arrival, the accuracy and precision decreases in case of multipath fading environment. Therefore, this chapter proposes an angle of signal propagation-based method where each node emits only two signals in a particular direction and knows its approximate position while receiving the second signal. Further, a method is proposed to define the coordinates of the nodes in reference to a local coordinate frame. The proposed method does the work with a smaller number of transmissions in the network even in the presence of malicious adversaries.

## Chapter 6

*Nisha Kandhoul, Netaji Subhas University of Technology, India*
*Sanjay K. Dhurandher, Netaji Subhas University of Technology, India*

Internet of Things(IoT) is a technical revolution of the internet where users, computing systems, and daily objects having sensing abilities, collaborate to provide innovative services in several application domains. Opportunistic IoT(OppIoT)is an extension of the opportunistic networks that exploits the interactions between the human-based communities and the IoT devices to increase the network connectivity and reliability. In this context, the security and privacy requirements play a crucial role as the collected information is exposed to a wide unknown audience. An adaptable infrastructure is required to handle the intrinsic vulnerabilities of OppIoT devices, with limited resources and heterogeneous technologies. This chapter elaborates the security requirements, the possible threats, and the current work conducted in the field of security in OppIoT networks.

## Chapter 7

*Vikash, Indian Institute of Information Technology, Allahabad, India*
*Lalita Mishra, Indian Institute of Information Technology, Allahabad, India*
*Shirshu Varma, Indian Institute of Information Technology, Allahabad, India*

Internet of things is one of the most rapidly growing research areas. Nowadays, IoT is applicable in various diverse areas because of its basic feature i.e., anything would be available to anyone at anytime. Further, IoT aims to provide service in a pervasive environment, although different problems crop up when the researchers move towards pervasiveness. Security and Privacy are the most intense problems in the field of IoT. There are various approaches available to handle these issues: Architectural security, Database security, Secure communication, and Middleware approaches. This chapter's authors concentrate on middleware approach from the security and privacy perceptive. Middleware can provide security by separating the end user from the actual complex system. Middleware also hides the actual complexity of the system from the user. So, the user will get the seamless services with no threats to security or privacy. This chapter provides a brief overview of secure middlewares and suggests the current research gaps as future directions.

Distributed denial of service (DDoS) attack is one of the most disastrous attacks that compromises the resources and services of the server. DDoS attack makes the services unavailable for its legitimate users by flooding the network with illegitimate traffic. Most commonly, it targets the bandwidth and resources of the server. This chapter discusses various types of DDoS attacks with their behavior. It describes the state-of-the-art of DDoS attacks. An emerging technology named "Software-defined networking" (SDN) has been developed for new generation networks. It has become a trending way of networking. Due to the centralized networking technology, SDN suffers from DDoS attacks. SDN controller manages the functionality of the complete network. Therefore, it is the most vulnerable target of the attackers to be attacked. This work illustrates how DDoS attacks affect the whole working of SDN. The objective of this chapter is also to provide a better understanding of DDoS attacks and how machine learning approaches may be used for detecting DDoS attacks.

As the lot of data is getting generated and captured in Internet of Things (IoT)—based industrial devices which is real time and unstructured in nature. The IoT technology—based sensors are the effective solution for monitoring these industrial processes in an efficient way. However, the real—time data storage and its processing in IoT applications is still a big challenge. This chapter proposes a new big data pipeline solution for storing and processing IoT sensor data. The proposed big data processing platform uses Apache Flume for efficiently collecting and transferring large amounts of IoT data from Cloud—based server into Hadoop Distributed File System for storage of IoT—based sensor data. Apache Storm is to be used for processing this real—time data. Next, the authors propose the use of hybrid prediction model of Density-based spatial clustering of applications with noise (DBSCAN) to remove sensor data outliers and provide better accuracy fault detection in IoT Industrial processes by using Support Vector Machine (SVM) machine learning classification technique.

## Chapter 10
*Bhanu Chander, Pondicherry University, India*

The Internet of Things (IoT) pictures an entire connected world, where things or devices are proficient to exchange a few measured data words and interrelate with additional things. This turns for a feasible digital demonstration of the existent world. Nonetheless, nearly all IoT things are simple to mistreat or compromise. Moreover, IoT devices are restricted in computation, power, and storage, so they are more vulnerable to bugs and attacks than endpoint devices like smartphones, tablets, and computers. Blockchain has remarkable interest from academics and industry because of its salient features including reduced dependencies on third parties, cryptographic security, immutability, decentralized nature, distributed nature, and anonymity. In the current scenario, blockchain with its features provides an anonymous framework for IoT. This chapter produces comprehensive knowledge of IoTs, Blockchain knowledge, security issues, Blockchain integration with IoT (BIoT), consensus, mining, message validation mechanisms, challenges, a solution, and future directions.

## Chapter 11
*Pratik Shrivastava, Madan Mohan Malaviya University of Technology, India*
*Udai Shanker, Madan Mohan Malaviya University of Technology, India*

Security in replicated distributed real time database system (RDRTDBS) is still explorative and, despite an increase in real-time applications, many issues and challenges remain in designing a more secure system model. However, very little research has been reported for maintaining security, timeliness, and mutual consistency. This chapter proposes the secure system model for RDRTDBS which secures the system from malicious attack. To prevent the request/response from malicious attack, authors have extended the system model with a cryptographic algorithm. In the cryptographic algorithm, a key must be secretly known only to the sender and receiver. Thus, in this chapter, authors have used the key generation algorithm to generate a key using an image. This secure system model maintains the confidentiality of the replicated data item and preserves its data integrity. It performs better in terms of malicious attack compared to other non-secure system models.

# Foreword

The Internet of Things (IoT)is probably one of the most important technological revolutions enabling anything, anywhere, anytime, and anywayconnectivity to all the things and probably all the people. The paradigm of IoT embracing sensor networks offers superscale connectivity of identifiable things comprising from across all walks of life. This almost limitless horizon makes the platform of sensor networks and IoT highly vulnerable to security threats and privacy breach.

This book indeed provides very useful insights of the security and privacy issues in sensor networks and IoT.

Chapter 1begins with detailed insights of security aspects in wireless mesh networks. Interesting reference has been provided onto digital signature based intrusion detection systems. This chapter motivates the readers to explore deep learning techniques and novel intrusion detection methods for securing wireless mesh networks.

Industry 4.0 is envisaged to transform the manufacturing units as smart factory through IIoT. The authors in chapter 2 have very beautifully presented the interesting material on IIoT adoption in manufacturing sector. Chapter 3 delves into security challenges in IoT systems. The authors have been successful in emphasizing the machine learning based solutions to address the security issues in IoT. Body Area Networks being a critical category of data networks are studied in chapter 4. The details of modeling the propagation channel in these networks offers very important material for readers interested in the area of body area networks. Node localization plays vital role in WSN implementation. Chapter 5 introduces an energy efficient scheme relying on computation but not on communication for localization in WSNs. Those working in the area of WSNs, will find this material very interesting to advance their research work. OppNets based on mobile pervasive wireless infrastructure are being evolved with self-organizing networking capabilities. Chapter 6 on OppNets offers advanced research material on security aspects in Opportunistic IoT. Middleware approach towards enhancing the security and privacy in IoT is presented in chapter 7. The authors have identified some research gaps in this area which can be very useful for those intended to pursue research in this specific domain. For a person entrusted with the responsibility of designing security solutions for data networks, it is essential to have comprehensive understanding of the DDoS attacks. Chapter 8 offers comprehensive material on this very critical aspect of securing the networks. The concept of "Big Data" hasindeed occupied one of the most important places in ICT. Chapter 9 introduces this topic with reference to storage and processing model in IoT scenario. Novel big data pipeline solution for storing and processing of IoT sensor data offers a very interesting reading material for the readers.

The book provides thoughtful and interesting reference to students, researchers and industry professionals as well working in the area of security and privacy of data networks. Blockchain technology operating in decentralized way is supposed to offer cost effective solutions for securing intelligent communication. Chapter 10 provides interesting information on the integration of this wonderful and disruptive technology with IoT. Finally, chapter 11 introduces the readers with secure system model for replicated distributed real time data base system. Provisioning of the security in these systems is relatively novel research area and the contents in the chapter are good enough to evince the interest for developing innovative solutions to the open technical challenges.

Securing the information networks and protecting the privacy of the users will continue to be important and critical for the network designers and a matter of concern for the end users. This book has presented critical analysis of the technical challenges related to provisioning of the security and privacy in the broader range of advanced networking technologies.

I appreciate the efforts put in by the authors for bringing out recent trends in the thematic area of security and privacy diligently mentioned and beautifully illustrated in the book.

Those interested in venturing out and acquainting themselves with theoretical analysis of security frameworks and recent developments in the domain of security and privacy aspects in a class of networks, this book is one of the best places to begin with.

Through my close interaction with, for so many years, I have observed keen interest and passion of Prof. Mayank Dave and Dr. Priyanka Ahlawatabout teaching and research in the area of data networks and its different dimensions. I am indeed delighted to see their domain knowledge being fructified in the form of this book. Their understanding and sound knowledge of the subject matter has indeed made it easy to present the text in simple and lucid style which can be used as a text book for students and a reference guide for researchers. I wish them all the very best in their future endeavors.

Happy reading.

*Brahmjit Singh*
*National Institute of Technology Kurukshetra-136119, India*

# Preface

The applications of sensor networks and internet of things (IoT) are becoming ubiquitous and are becoming important for industrial and home applications. Sensor networks and IoT have applications in different areas such as smart cities. Rapid research and technological developments occurring in the area require updated knowledge for the researchers and application developers. However, threats and vulnerabilities make the applications in sensor networks and IoT domain critical as IoT has already started altering our world. It is expected that 21 million devices will be connected to the Internet by 2021, and about 16 million of them will be part of IoT. IoT makes use of interconnecting devices with the internet and thus providing an environment and possibility of creating applications that provide positive impact on human lives. IoT is going play a very important role in cyber physical systems and critical infrastructure. The huge number of IoT devices is posing significant challenges for IoT security like authentication and privacy preserving algorithms, distributed denial of service (DDoS) attacks, authorization, authentication, confidentiality and integrity of data, which need to be seriously addressed. This book is meant to provide a platform for researchers, industry professions and academicians to present the state-of-the-art developments, challenges and unsolved open problems in the field of security and privacy in sensor networks and IoT.

The purpose of this book is to provide relevant theoretical framework and latest research trends in Security and Privacy issues in Sensor Networks and IoT. The book will also provide insight to the problem-solving abilities, tools and techniques. The target audience of this book is students, researchers and professionals working in this domain or want to gain insight into this important area. Each chapter of this book contributes towards providing knowledge in the diverse domain of the area.

The book is a compilation of eleven chapters comprising of chapters from different sub domains collectively presenting and addressing issues, challenges and solutions related to security and privacy in sensor networks and IoT. The book begins by understanding the problems in wireless mesh networks that occur due to malicious attacks, their detection and prevention. Other chapters look into other important aspects of security and privacy namely, Industrial IoTs (IIoTs), machine learning for IoT security, securing communication channel, routing and middleware for IoT. DDoS are attacks that effect the network in big way as these attacks make the network unable to deliver the services desired by the application users.

Software defined networking is a recent paradigm that is oriented towards reducing the cost of networks while improving the efficiency and maintainability. Big data storage and processing needed for IoT data require special attention for industrial applications. Blockchain technology has gained a lot of attention in recent years and their integration into IoT applications means considering distributed architectures for application service. Real time database system (RTDBS) are important for IoT applications in smart cities and thus securing RTDBS from malicious attacks require new protocols. The last chapter of the book focuses on providing a new security model for RTDBS.

A brief description of the chapters of this book is as follows:

**Chapter 1:** This chapter aims to review and identify the different challenges, architecture, and protocols designed for wireless mesh networks security. Due to suppression of central administration in WMN, network functioning like network controls, management, routing, switching, packet forwarding etc. are distributed among nodes itself, either collectively or individually. So, cooperation among nodes is highly solicited. The nodes may misbehave by refusing to provide service or dropping down the packets because of its selfishness and malicious activity. The identification of misbehaving nodes and prevention from them is one of the biggest challenges. Hence, the prime target of the chapter is to provide an overview of existing intrusion detection and prevention approaches, secure routing or framework that can recognize and prevent from the malicious activities. The digital signature-based intrusion detection system (IDS) offers secure acknowledgment and the authentication mechanism has also been discussed. The expectation is that the digital signature-based IDS will overcome the weakness of existing IDS. The digital signature-based IDS nowadays receives a lot of scrutiny for enhancing the security issues in WMNs. Earlier systems used different kinds of intrusion detection and prevention schemes with other valid authentication systems to keep the network free from malicious attacks of intruders. However, the modern IoT networks are comprised of nodes that have open characteristics with a limited survival life time. Hence, there is a need of such a security system which will provide safety at all the important components that form the basic WMNs. The system that identifies the malicious attacks and then tries to provide the complete safety against such attacks in WMNs with overcoming problems of reputation, receiver collision, false identity problems, etc., is digital signature-based IDS. To solve these problems, deep learning techniques and customized IDS with new approaches and digital signature-based credentials are used to provide a fundamental solution for it.

**Chapter 2:** This book chapter presents a comprehensive view of industrial automation using IIoT. Today, advanced industries are leading in a new age of physical production backed by the information-based economy. It is mainly the convergence of industrial systems with advanced near real-time computing and analytics, powered by low cost and low power sensing devices leveraging global internet connectivity. The key benefits of IIoT systems are also discussed in this chapter. The widespread proliferation of IIoT is being enabled by availability and affordability of sensors, computing resources (e.g. processors) and technologies like cloud computing, big data and machine learning backed by fast and reliable networks. IIoT, helps industries in improving their productivity and operational efficiency, enhancing worker safety etc. IIoT also has the capability to create new revenue streams by transforming organizations from being product-centric to customer-centric. The computing environment facilitates capture of and access to real-time information for decision making. IIoT adoption is fast emerging as a key driver in transforming and automating manufacturing processes. As with any large system, IIoT solutions need to be protected both physically and for cyber threats and should be continuously monitored for vulnerabilities. For large scale global businesses, journey towards adoption of IIoT really translates into three major aspects like creating new Industry Model, harnessing and capitalizing the value of data through descriptive, predictive and prescriptive analytics and building talent and workforce ready for the future.

**Chapter 3:** This chapter provides various machine learning techniques to mitigate security attacks in IoT. Energy conservation, memory, and computational speed plays an essential part in these smart devices whereas on the other hand, these smarter things are open to wide range of security attack challenges. There are still major concerns lurking around the IoT ecosystem due to security threats. Major IoT security concerns are DoS, sensitive data exposure, unauthorized device access, etc. The main motivation of this chapter is to brief all the security issues existing in the Internet of Things (IoT) along with an analysis of the privacy issues. The chapter mainly focuses on the security loopholes arising out of the information exchange technologies used in IoT. It further discusses security solutions based on machine learning (ML) techniques including supervised learning, unsupervised learning and reinforcement learning. The ubiquitous nature of IoT and device characteristics make the network more prone to security threats and attacks. Depending on the data being communicated over the network, it inhibits an interest over the attackers with a wide range of privacy exploration. ML techniques are likely to provides solutions for the various concerns like privacy of data, data reliability, correct responses from the connected devices, trust-worthy devices and autonomous recovery of the device when compromised.

**Chapter 4:** This chapter analyzes the effect of channel modeling on the intercept behavior of a wireless body area network (BAN) while taking optimal sensor scheduling into account. BAN provides various applications using wearable devices. BAN is a type of personal area networks (PAN). BANs form a significant part of health care applications due to increase in medical expenses. Moreover, the patients suffering from chronic diseases who require only restorative observation need not be admitted to a hospital as this can be easily done via intelligent monitoring of the patient's body using wearable devices. According to Moore's law, the development of small and handheld devices which could be used for communication around human bodies was certain. Wearable computing is an interesting application such as physiological/medical monitoring of temperature, heart rate, and blood pressure, etc. BANs are basically intended to be used for health care applications where long-lasting and reliable operation is a must. To elongate the life of a BAN, low power sensors with a short range are employed. Other methods taken into account for long term usage of a BAN is communication using relays, controlling transmitting power of a sensor and adapting the link between sensors accordingly. Due to basic nature of wireless medium and the radio wave propagation's broadcast nature, the wireless communication can be overheard by the eavesdropper. To safeguard the BAN, the propagation channels need to be characterized and modeled for designing reliable communication systems.

**Chapter 5:** In wireless sensor networks (WSNs), secure localization is helpful in reporting an event's origin, assisting querying of sensors, routing, and various cyber-physical system applications, where sensors are required to report geographically meaningful data for location-based applications. One of the accurate ways of localization is the use of anchor nodes which are generally equipped with the global positioning system. However, in range-based approaches used in literature, like angle of arrival, the accuracy and precision decreases in case of multipath fading environment. Therefore, this chapter proposes an angle of signal propagation-based method where each node emits only two signals in a particular direction and knows its approximate position while receiving the second signal. Further, a method is proposed to define the coordinates of the nodes in reference to a local coordinate frame. The proposed method does the work with a smaller number of transmissions in the network even in the presence of malicious adversaries. This proposed scheme relies more on computation rather than communication. This scheme is expected to be much more energy-efficient scheme compared to other methods used previously for localization in WSNs.

**Chapter 6:** With the increased connectivity to the internet and the technical revolution where everyone is surrounded by sensors, ensuring the security of the users and the messages exchanged by them is a major challenge for Opportunistic IoT (OppIoT) networks. The user data is continuously being sensed and without their knowledge and active participation, the users' privacy is at stake. So, the requirement is to design secure routing protocols that can be easily implemented across varying technologies and platforms like OppIoT that comprise of a wide spectrum of devices and human users. These security enhancement techniques must aim at mitigating as many attacks as possible, while achieving good performance in routing. The security techniques applied should not degrade the performance of the network and must consider the limited power and buffer availability in these devices. There is a huge scope for research in the field of security of OppIoT networks as not much work has yet been conducted in this field. OppIoT are may be considered as an extension of the opportunistic networks, which exploit the interactions between the human based communities and the IoT devices to increase the network connectivity and reliability. An adaptable infrastructure is required to handle the intrinsic vulnerabilities of OppIoT devices, with limited resources and heterogeneous technologies. This chapter elaborates the security requirements, the possible threats and the current work conducted in the field of security in OppIoT networks

**Chapter 7:** This chapter presents middleware approach to enhance the security and privacy in IoT. IoT aims to provide service in pervasive environment. Different problems emerge when the researchers move towards pervasiveness. Security and privacy are the most intense problems in the field of IoT. There are various approaches available to handle these issues, which are categorized as architectural security, database security, secure communication, and middleware approaches. However, the authors concentrate on middleware approach from security and privacy perceptive. Middleware can provide security by separating the end user with the actual complex system. Additionally, middleware hides the actual complexity of the system from the user. So, the user will get the seamless services without any threat of security and privacy. Middleware is a powerful tool to the ease of application development and deployment that can perform task in complex or critical environments. The middleware framework can facilitate new features and provide benefits to security and privacy of IoT. It is helpful to decide the research direction with the goal of security and privacy for IoT along with validation and verification. The major outcome of this chapter is the concept of middleware solution to secure IoT, which is available to wider community of academics and practitioners and application areas. The chapter reviews different middleware according to security and privacy challenges. Some clear research gaps are also identified as future directions.

**Chapter 8:** This chapter presents brief review of DDoS attacks and its defense mechanisms using machine learning techniques-based IDS. These attacks have become one of the most vulnerable attacks for the networks. They are becoming stronger with the advancement of the network technologies, which make suffered the legitimate users for the network services. Some efficient solutions should be developed for the detection and mitigation of these vulnerable attacks. It makes the services unavailable for its legitimate users by flooding the network with illegitimate traffic. Most commonly, it targets the bandwidth and resources of the server. This chapter discusses various types of DDoS attacks with their behavior. It describes the state-of-the-art of DDoS attacks. An emerging technology named "software-defined networking" (SDN) has been developed for new generation networks. It has become a trending way of networking. Due to the centralized networking technology, SDN suffers from DDoS attacks. SDN controller manages the functionality of the complete network. Therefore, it is the most vulnerable target of the attackers to be attacked. This work illustrates how DDoS attacks affect the whole working of SDN. The objective of this chapter is also to provide a better understanding of DDoS attacks and how machine learning approaches may be used for detecting DDoS attacks. Different layers of the SDN architecture and impacts of DDoS attacks have also been presented in the chapter. It also includes a discussion of required steps in an IDS that is based on machine learning classification. Machine learning-based IDS have better prediction capability because of having training of the data. There is a requirement of developing efficient machine learning based defense solutions for the real-time network traffic. Finally, in the chapter some machine learning based solutions for DDoS detection in SDN are also presented.

**Chapter 9:** This chapter presents an efficient big data-based storage and processing model in IoT for improving accuracy fault detection in industrial processes. In this chapter, a scheme to solve the challenge of real time processing of sensor-based data and fault detection of industrial assembly data is presented. The IoT based sensor data is stored on Hadoop Distributed File System (HDFS), which is processed with Apache Storm on Big Data processing pipeline. The IoT technology-based sensors is the effective solution for monitoring these industrial processes in efficient way. However, the real time data storage and its processing in IoT applications is still a big challenge. The authors propose a new big data pipeline solution for storing and processing IoT sensor data. The solution uses Apache Flume for efficiently collecting and transferring large amounts of IoT data from Cloud based server into HDFS for data storage. The hybrid prediction model of density-based spatial clustering of applications with noise (DBSCAN) is used to remove sensor data outliers. Support Vector Means (SVM) is used for classification. The results suggest that this hybrid prediction model is scalable for data processing of IoT based sensor data and for accurate detection of faults than the traditional models.

**Chapter 10:** This chapter describes blockchain technology integration in IOT and applications. As IoT pictures entire connected world, where things or devices are proficient to exchange a few measured data words in addition interrelate with additional things. This turns for a feasible digital demonstration of the existent world. Nonetheless, nearly all IoT things are simple to mistreat or compromise. Moreover, IoT devices are restricted in computation, power, storage, so they are more defenseless to bugs, attacks than endpoint devices like smartphone's, tablets and computers, etc. The purpose of the blockchain concept has enlarged ahead of it's utilizing for Bitcoin production as well as transaction dealings. At present, blockchain along with modifications are utilized to protect any type of transactions whether human-to-human or machine-to-machine connections. Blockchain adoption appears to be revolutionizing the IoT. This chapter inspects the up-to-date of blockchain-IoT (BIoT) equipment and future applications in areas like big-data, healthcare, financial services, agriculture, and energy management. Moreover, BIoT integration issues, challenges, applications, mining, and message validation methods are also explained. Blockchain has acknowledged remarkable interest from academe and industry because of its salient features including reduces dependencies on third parties, cryptographically secured, immutability, decentralization, distributed and anonymity. In the current scenario, blockchain with its above-mentioned features provides an anonymous framework for IoT. This chapter also provides blockchain integration with IoT (BIoT), mechanisms, challenges, solution, and future directions related to this upcoming technology.

**Chapter 11:** The chapter aims to present a secure system model for replicated Distributed real time database system (DRTDBS). The confluence of real time systems, communication network, and database systems is creating a RTDBS. Recently, the demand of RTDBS is expanding rapidly. A large number of real time applications such as stock management system, banking system, business information system, and air traffic control system generate massive amount of data. With IoT such applications are likely to increase. A DRTDBS is designed to satisfy the timeliness demand of real time traffic (RTT) such that temporal consistency of such huge amount of data can be maintained. The research efforts are being made in the concurrency control protocol (CCP), commit protocol (CP), replication technique (RT) and buffer management (BM) to maximize majority of RTTs to get successfully complete within their deadline. Security in replicated distributed real time database system (RDRTDBS) is still explorative and, despite an increase in real time applications, many issues and challenges remain in designing a more secure system model. However, very little research has been reported for maintaining security, timeliness, and mutual consistency. The aim of this chapter is to propose the secure system model for RDRTDBS that secures the system from the malicious attacks. To prevent the request/response from malicious attack, we have extended the system model with cryptographic algorithm. In the cryptographic algorithm, a key must be secretly known only to the sender and receiver. Thus, in this chapter, we have used the key generation algorithm that generates the key using an image. This secure system model maintains the confidentiality of the replicated data item and preserves its data integrity. It performs better in terms of malicious attack compare to other non-secure system models.

The book is aimed to be an important reference source developed with current literature in the field of security and privacy in sensor networks and IoT. It thus, provides further research opportunities that can be the primary and major resources necessary for researchers, academicians, students, faculties, and scientists, across the globe, to understand and develop new inventions in the area of sensor networks and IoT security.

# Acknowledgment

First of all, we express our deepest gratitude to the Almighty, who blessed us with the enthusiasm and interest to complete the book on the Security and Privacy Issues in Sensor Networks and IoT successfully. We are deeply indebted for the time spent by the reviewers by providing their expertise in reviewing different papers submitted for inclusion in this book. We highly appreciate their efforts in timely completion of this project. We profusely thank our reviewers of each chapter of this book for their outstanding wisdom, vision, guidance and enthusiastic involvement.

We would like to thank the editorial advisory board regarding the improvement of quality, coherence and the content presentation of this book. We specially mention efforts of IGI Global team Mrs. Lindsay Wertman (née Johnston), Managing Director, IGI Global; Ms. Maria Rohde; Ms. Carlee Nilphai; Mrs. Jordan Tepper for their constant encouragement, continuous assistance, technical support and also ensuring the progress the book in right direction. We are also thankful to faculty of Department of Computer Engineering, National Institute of Technology, Kurukshetra for their constructive comments and guidance. We are extremely thankful to Ms. Sonam Bhardwaj for helping us in editing and make it possible not to miss any important deadline.

We are very thankful to our honorable Director, National Institute of Technology, Kurukshetra to provide us all kind of supports and facilities which have been immensely helpful for the completion of this book.

*Mayank Dave*
*National Institute of Technology, Kurukshetra, India*

*Priyanka Ahlawat*
*National Institute of Technology, Kurukshetra, India*

# Chapter 1
# Wireless Mesh Network Security, Architecture, and Protocols

**Sachin Kumar Gupta**
*Shri Mata Vaishno Devi University, India*

**Aabid Rashid Wani**
*Shri Mata Vaishno Devi University, India*

**Santosh Kumar**
iD https://orcid.org/0000-0003-2264-9014
*Department of Computer Science and Engineering, Dr. SPM IIIT Naya Raipur, Chhattisgarh, India*

**Ashutosh Srivastava**
*RST Ecoenergy Private Limited, Mirzapur, India*

**Diwankshi Sharma**
*Shri Mata Vaishno Devi University, India*

## ABSTRACT

*Due to suppression of central administration in WMN, network functioning like network controls, management, routing, switching, packet forwarding etc. are distributed among nodes, either collectively or individually. So, cooperation among nodes is highly solicited. However, there may exist node's malicious activities because of its open characteristics and limited available battery power. The nodes may misbehave by refusing to provide service or dropping down the packets because of its selfishness and malicious activity. The identification of misbehaving nodes and prevention from them can be one of the biggest challenges. Hence, the prime target of the chapter is to provide an overview of existing intrusion detection and prevention approaches,*

DOI: 10.4018/978-1-7998-0373-7.ch001

*and secure routing or framework that can recognize and prevent from the malicious activities. The digital signature-based IDS to offer secure acknowledgment and an authentication mechanism has also been discussed. The expectation is the digital signature-based IDS will overcome the weakness of existing IDS.*

## INTRODUCTION

The revolution in communication technology has made the life of a common person too much easy. Because of this revolution, a lot of fields came into the market. For example, the wireless connectivity of mobile users all over the world, the real-time communications, internet of things (IoT), etc. These modern facilities are only due to the technologies like internet, cellular, IEEE 802.11, IEEE 802.15, IEEE 802.16, sensor networks, etc. Among all these revolutionary technologies, the most important wireless technology named Wireless Mesh Networks (WMNs) that provide facilities to the great extent in day to day life of common person. The facilities are in terms of implementations like Private, local, campus, urban regions, etc. The cause behind these applications is the nature of nodes using in deploying wireless mesh networks. As these nodes are dynamic, self-organized, self-configurable, self-healing, low-cost, easy maintenance. Among these characteristics of nodes deployed for establishing WMNs, the dynamic topology and the multi-hop nature leads WMNs vulnerabilities to security attacks. So, the network should consider security issues like authenticity of network traffic (free from masquerading of nodes), non- repudiation, authorization among users, anonymity, access control and secure routing etc. These security issues in WMNs can be achieved with some important key management schemes. These include intrusion prevention (Offense against fraudulent nodes, as well as for authentication and encryption), intrusion detection (Ideal for preventing the invasion or reducing the harm), and intrusion responses. Hence, the security-based WMNs overcome the threats from external or internal intruders (attackers) (Sgora et al., 2016).

It has been come for the long time that wireless networks should be developed and deployed efficiently. The necessity of efficient development and deployment was to fulfil our application requirements. That is the requirements needed for each application like throughput, wait time, memory, safety etc. must be eOncountered. So, the researchers always remain behind to create new dedicated wireless technologies and standards. Keeping in mind that these technologies should met our above application requirements. Due to which number of modern technologies took birth viz IEEE 802.11(Wi Fi), IEEE 802.15 (including Bluetooth and ZigBee), and IEEE 802.16 (WiMAX). These technologies were so much important due to their affordable capabilities and wireless mesh capacities.

Since WMNs are inherently unordered networks. Such unstructured/unordered nature of the network is due to high mobility and frequent changing of topology with time. The unorganized nature of WMNs leads to hazy packet forwarding from source to destination within the same network. So to establish ordered/structured nature of packet forwarding within the WMNs. We need requirement of necessary routing protocols (DSR, AODV, OLSR etc.) that not only establishes optimized routes with non–neighbouring nodes but also accountable for exploration, setting up and maintaining of such routes. Hence, routing protocols not only optimize routes but also brought optimization in many important parameters of WMN. For example, the optimization in hop count (minimum hop count), optimized delay, optimization in power utilization, etc. The optimization in all these parameters results in increased data rate with maximum throughputs.

In WMNs as many authentication algorithms are using to solve the challenges of efficient routing phenomenon of nodes and mutual authentication among nodes. Hence the researchers always tried their best to create such a security system algorithm that can make the wireless network system secure at its peak. But it has been not achieved till now as intruders/enemy always come with new attacks to the given wireless networks. So new security systems came into market which solve their given challenges as per their inbuilt algorithms. Among all the known security algorithms, the one which is using at its peak nowadays is intrusion detection and prevention system. The Intrusion Detection System (IDS) and Intrusion Prevention System (IPS) not only detect the inclusion of intruders/attackers into the given network but also prevents their malicious attacks to the great extent. The IDS can be implemented at the router for traffic analysis or can be implemented at the base controller that controls the given WMNs. The intrusion detection identifies the given attacks on anomaly based (behavioural nature of nodes) or from known pattern of threats (signature-based).From above it has been cleared that research is going at its best to solve the real challenges of routing problems with mutual authentications of nodes for secure communications in WMNs. But still there is need of lot of research as per security point of view as the attackers/intruders learn new techniques of hacking the given wireless networks. Hence, more than one authentication schemes with necessary protocols become a trend to explore the WMNs to provide all essential/ potential services to the mankind for example internet to remote areas, health related applications, peer to peer communication etc.

## Motivation and Contribution

The digital signature-based IDS nowadays receiving lot of scrutiny for enhancing the security issues in WMNs. Before digital signature-based IDS different kinds of intrusion detection and prevention schemes with other valid authentication systems

3

were developed for keeping the same network free from malicious attacks of intruders. But due to open characteristic nature of nodes with limited survival lifetime. The security against vulnerable attacks cannot be achieved up to the user's satisfactions. So there are maximum chances of drooping down the packets and easy stealing of the valid data exchanging between authorized users by non-authorized attackers. In these networks, the intruders can attack at any place like client or router or server or packet level for accessing the useful information. Hence, there is a need of such a security system which will provide safety at all the important components that form the basic WMNs. The security system that identifies the malicious attacks and then tries to provide the complete safety against such attacks in WMNs with overcoming problems of reputation, receiver collision, false identity problems, etc. is digital signature-based IDS. Hence, the digital signature-based IDS may establish satisfied communication among the users using WMNs for accessing their connectivity in real-time communications in this digital modern world.

## LITERATURE: REPORTED WORKS AND BACKGROUND

WMS is considered the promising technology for the future era of broad modern services. This technology not only comes up with most economical services in the field of health and medicines. But also provide internet access to remote areas as well as in natural disaster areas. This mesh-based network was prone to assaults on security and routing overheads due to dynamic topology and multi hop nature. No doubt the security challenges because of these constraints were accomplished to the target value. And the result leads to information confidentiality, resource heterogeneity, network traffic authentication, and user authorization etc. to the achievable value (Feng et al., 2008).

Several researchers have done the qualitative and quantitative analysis to point the various kind of attacks and come up with its solutions in WMN. Among them, few most relevant to this article are as below.

(Yong et al, 2008). In this work the security goals were carried out only at the different layers of WMS (to achieve mutual authentication between nodes) excluding the mesh user's privacy. Hence, there is a need of that security system which provides not only authentication among nodes but also preserve effectively mesh user's privacy. This challenge was solved by new privacy scheme named Enhanced Authentication scheme. The Enhanced Authentication scheme is anonymous authentication scheme based on CPK (Combined Public Key) and Blind Signature in the elliptical curve domain. The CPK and Blind Signature key managements provide authentication not only between Mesh Clients (MCs) and Mesh Routers (MRS). But also accomplished privacy among MCs.

(Zhijun et al., 2008) reported that the Wireless Sensor Nodes (WSN) face many problems viz, low potential for computing, limited storage, finite energy resources, lack of infrastructure, and the big constraint that is susceptibility to physical attacks (unsecure network). So, the data that is gathered and transferred by the sensor nodes within the WSN to particular destination must follow confidentiality, integrity, entity authentication, availability etc. To achieve these security goals key distribution and management comes into play. In key distribution and management many approaches came into market one after another to make WSN more secure. These approaches are Straight forward Approaches(single network-wide key, fully pairwise keys scheme, Kerberos-like key distribution), Schemes based on Initial Trust Model, Basic Random Probabilistic Key Distribution Scheme, Enhancement of Random Key Distribution, Key Distribution Using Combinatorial Design, Group Key Distribution, Public Key Feasibility. General Intrusion Detection and Intrusion Tolerance, Authentication, privacy and many more because of many innovative protocols and security techniques. The WSN became free from adversary attacks to the great extent. But still more future research is required to make WSN fully secure.

(Trong, et al., 2013) evaluate the routing performance of 802.11 technologies based on WMS. As 802.11 based WMNs are frequently using in establishing the real time communications. So, the nodes forming such networks come into intruder attacks easily. The situation becomes more critical at that time when these are deploying in an ad-hoc fashion. As in ad-hoc network each node should be self-configurable, self-healer and should perform routing of packets efficiently within the network. So, every node must behave as normal node but not as misbehaving node. As misbehaving nodes increases average end to end delay and average packet dropping (bits/sec). Hence, this research studies performance of two routing protocols known Optimized Link State Routing (OLSR) and Ad hoc On-demand Distance Vector (AODV) under variable number of intruder attacks. It was cleared that OLST routing protocol shows better performance in terms of above parameters than AODV under adversary attacks.

(Hector et al., 2011) reported that WMN is not appropriate for every type of application. Also, the mesh network is not capable to integrate all the available wireless technologies as lack of ability to communicate among themselves. To fix such a problem SORA (Service Oriented Routing Protocol) was developed by them. The SORA is a unified routing algorithm that overcomes various security challenges. It was implemented with three security techniques of key management, key shifting, and node relation monitoring. Hence, with the unification of different technologies confidentiality, integrity, and authentication also made possible by implementing this technique.

(Gaurav et al., 2018) stated that due to the vast application domain of networked UAVs like in medical field, video and photography (real-time analysis), environmental issues, law enforcement surveillance, public safety communication etc. The UAVs still facing real-time challenges/problems, for example malicious threats over open-air radio environment (security issue), a spectrum sharing, deployment and route scheduling problem. Among these authors focused on security issues that is implementation of IDS (Intrusion Detection System) mechanism with networked UAVs. The IDS deals with attacks and vulnerabilities over networked UAV domain. Hence the IDS watches confidentiality, integrity, availability, and authenticity to make the networked UAV more secured. So, IDS converges both effectiveness (Lowest reporting failures with low disruptions of service) and efficiency (reducing processing and communication overhead) in terms of both security and performance.

(Ping et al., 2009) reported the different security issues concerning WMN. WMN needs additional committed security solution then old-fashioned security scheme. At first, the article presents the number of possible threats to security in WMS. Afterward try to introduce possible solutions for that problem like key management, routing security, intrusion detection. In the end, they also do the comparative study between various solutions of these problems in terms of their merit and drawback. And finally, come up with remaining challenges in the area. WMN security architecture is a fast developing and promising field in the arena of wireless networking. It is required to put more effort in following sections like defence against DoS attacks, cross-layer security architecture, security protection for multicast, protection of traffic flow and location information privacy etc.

(Vural et al., 2013) this survey paper is based on experimental evaluation and observation that is experimented for the number of city where the mesh network has been deployed. To provide the Internet service across city, backbone infrastructure-based WMS has proven itself as a cost-effective technique. Their study evaluates the performance of city wise WMS deployments. Also, various potential research directions have been pointed out in this filed. They address deployment concerning aspects such as connectivity, transmission range, proper planning etc. and physical related attributes issues like pertaining to interference, path loss measurement, multipath effects etc. Apart from common wireless networking problems they also focus on users' behaviour, nature of traffic flow, and usage patterns emerge to guide someone for future research directions. Anyone can extract valuable information from this literature survey who wish to deploy WMS for ubiquitous Internet access in the metropolitan areas.

## KEY ASPECTS IN WMNS

WMNs terms wireless networks where wireless nodes are co-operative directly or indirectly with one or more peer nodes. Traditionally, the word mesh defines there is the connection between each node to each and every possible destination nodes directly. But as per contemporary definition, only a subset of nodes binds to one another. There are two kinds of nodes in WMNs: mesh routers and mesh customers. Both types of nodes play the function of the host and routers.

## Architecture

Based on the network topologies and the node's functionalities, WMNs architecture is classified into three different categories:

## Backbone Based WMNs

The backbone/infrastructure based WMN architecture is configured by linking various kind of mesh nodes that is shown in Figure 1. The network architecture comprises mesh routers that form an infrastructure for mesh customers that link to them, (Hamid et al., 2006). In this network, various kinds of radio technologies are used to create backbone, where technology differs significantly. Typically, two different kind of radios are used by routers, one for backbone communication and second one for

*Figure 1. Example of backbone based WMNs*

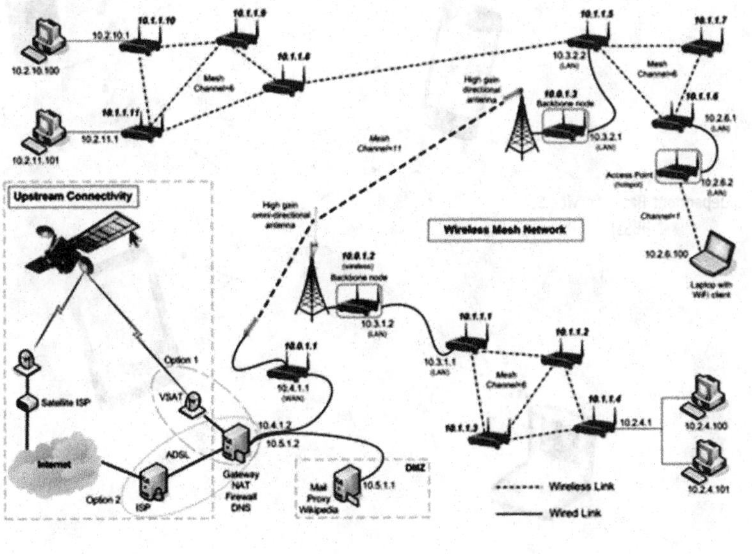

user communication. The backbone communication uses Directional antenna for long range transmission, whereas user communication is generally done through Omni-directional antenna. Mesh routers may be connected to distributed systems (i.e. Internet) and offers backbone for mesh clients, through gateway functionality, (Wang et al., 2006). With the help of Ethernet interface, clients may also connect to mesh routers through ethernet links. The packet may use several router hops to reach its end destination, (Asherson et al., n.d.) The simplicity is key merit of this network whereas its network scalability and high resource limitations are disadvantages.

## Clients Based WMNs

In this architecture, peer to peer connectivity is there among mesh clients (i.e. suppression of mesh routers), as illustrated in Figure 2. The client nodes are responsible for all kinds of routing and other configuration functionalities and provide connectivity to end users for various applications. The clients themselves execute various network responsibilities and maintain the network connectivity. The multiple intermediate clients are being used to deliver packet from sender to receiver. This architecture includes various advantages such as easy deployment, fast configurable, etc., and only suitable for indoor environment due to scalability problem (not fit for metropolitan level networks) (Hamid et al., 2006).

*Figure 2. Example of client based WMNs*

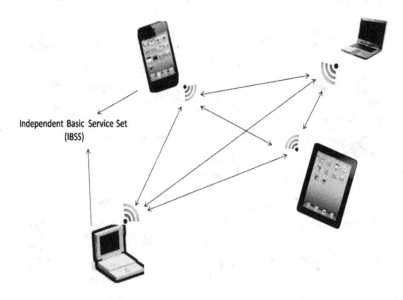

## Hybrid WMNs

Hybrid WMN comprises the features of both architecture backbone and client-based is shown in Figure 3 Any mesh customer can take assistance from both mesh routers as well as directly with other mesh customers in order to access the network. The backbone provides connectivity to other networks such as the Internet, Wi-Fi, Wimax, Cellular, Bluetooth WSN and so on. The routing capacities of the customers, on the other hand, enhance connectivity and coverage within the WMN. This architecture becomes very essential in modern terms and is most relevant in WMNs. The progress of WMNs lies on how it cooperates with other kinds of existing wireless networking solutions, (Hamid et al., 2006).

*Figure 3. Example of hybrid WMNs*

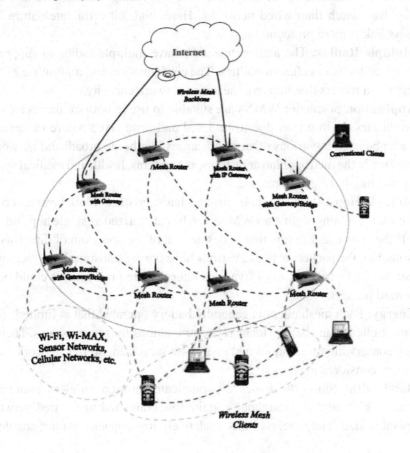

## Characteristics

The main characteristics of WMNs are as follow, (Hamid et al., 2009), (Wang, et al., 2006), (Asherson et al., n.d.; Salem et al., 2006), (Stajano et al., 2002), and (Stalling et al., 2006):

**Wireless Backbone or Infrastructure:** The WMNs are able to provide long coverage area, network connectivity, and robustness because of involvement of wireless backbone with mesh routers. Hence, end users get the reliable connection.

**Integration:** WMNs architecture can also integrate with dissimilar kind of networks like WSN, MANET, Bluetooth, Cellular, Internet, Wimax, Wi-Fi (a, b, g, n, etc.) that is possible through enabling of gateway and bridging functionalities.

**Mobility:** Generally, WMNs architecture has a fixed node so provide a relatively static environment in the network. But sometimes clients can be moveable that could generate lag in network convergence. Hence, topology can be dynamic.

**Fluctuating Link Capacity:** In WMNs, one end-to-end path or communication session can be completed by the multiple sessions. Channels through which wireless nodes are communicating are subjected to noise, fading, interference and have a smaller bandwidth than wired networks. Here, high bit error rate nature of the wireless link is more profound.

**Multiple Radios:** The architecture may have multiple radios to support and perform various access functionalities. The configuration and routing are executed among mesh routers that improve the entire network capacity.

**Application Scenario:** WMNs are suitable to use in both militaries as well as non-militaries environment due to its fixed and semi fixed nature of devices. It may also be used in various other applications like home broadband networking, networking in the metropolitan area, transport systems, health and medical systems, safety and monitoring systems, etc.

**Multiple Hops:** Based on their various characteristics of link layer and routing protocols, the routing algorithm WMNs can be categorized as single hop and multi-hop. If the source and destination nodes are out of the spectrum of direct wireless transmission, the packets may use various hops during transmission to achieve the destination node. In other words, various intermediate / relay nodes could be used to forward packets.

**Energy:** Each mesh client is generally battery operated that is limited. All the time mesh clients can't be used for heavy commutations and transmissions. Therefore, energy conservation techniques and energy-aware routing in this network become necessary considerations.

**Bandwidth:** Since the nodes communicate through wireless connections, this network's realized throughput is quite low compared to a wired network of comparable size. The wireless links ' relatively low capacity cannot enable real-time transmission. Therefore, the transmission of packets leads to delay. In addition, wireless connections are quite prone to errors, which can further degrade throughput owing to retransmission of the bottom layer, etc.

**Lightweight Terminals:** The mesh customers and routers usually have less ability for CPU processing, tiny memory size, and less ability for power storage. These devices therefore needed optimized algorithms and mechanisms to enforce computing and communication features.

## Security

### Challenges and Constraints

WMN can be revealed from some underlying threats which are common for both wired as well as wireless media. Therefore, the information of such networks may be changed, delayed, replayed, intercepted, or inserted by the new messages. However, such networks are more problematic to be completely secured for the following various reasons (Akyildiz et al., 2005), (Zhang et al., 2006):

**Nature of Multiple Hop:** This nature introduced the delays in the network in order to detect the threats and prevent these threats. In the literature, most of the security systems are suitable for single hop networks. Therefore, presented schemes are not enough sufficient to provide protection WMN from being attacked.

**Multitier System Security:** Nowadays, hybrid architecture is playing the significant role in society. Therefore, in such kind of networks, network security is not only needed among the mesh client, but also among mesh routers as well as between mesh clients and mesh routers.

**Multisystem Security:** Due to the development of WMNs to provide interoperability among various wireless systems such as WPAN, WI-FI, WIMAX, WSN, MANET, Cellular, Internet, etc. Security mechanisms thus become essential so that inter-network communications can be provided seamlessly without compromising safety across all networks.

Moreover, WMNs also suffering from various constraints as described in section 3.2, such as energy limitations, bandwidth limitations, less CPU, less memory, mobility scalability etc. (Siddique, et al., 2007), and (Goa et al., 2010).

### Issues

The security issues in WMNs are almost identical as in other communication technologies. The different security issues are as follow given below (Siddique et al., 2007), (Goa et al., 2010), (Malik et al., 2011), (Seyedzadegam et al., 2011), (Naveed et al., 2009), (Egners et al., 2010), (Ren et al., 2010), and (Durahim et al., 2010):

11

**Confidentiality:** Only communicating entities should understand the content of message those are authorised to access it (i.e. authorised parties).

**Integrity:** Ensure that when message is pumped into the network, the message that arrives at other party is the same as the message that was originally sent by sender (i.e. message should not be altered in any means).

**Availability:** The message must be available to the authorized entities despite the various attacks as described in just immediate section.

**Authentication:** Ensure that whosoever provides or accesses the sensitive message must be authorized parties. I.e. there must be some processes to recognize the WMN client to confirm that messages are really sent by authenticate client rather than by fabricated one.

**Authorization:** Before performing any task by WMN clients, there should be some mechanism from the trusted authority to ensure that corresponding clients have the right to do this task.

**Non-Repudiation:** It guarantees that the WMN sender and the receiver client cannot later deny that they have ever sent or received a signal.

**Accounting:** When a client uses any services, there must be some mechanism to measure how much resources have been consumed by the particular client for billing purpose or to maintain the consumption records.

**Fairness:** In WMN architecture, the Medium Access Control (MAC) sub-layer is responsible to ensure that no any client suffers from resources or bandwidth starvation. Moreover, the path selection or routing protocol is responsible to provide fairness on the forwarded traffic.

**Anonymity:** All the information which are used to recognize the particular owner or the client must be kept secure and must not be disclosed to other communicating clients.

## Attacks

The various security attacks in WMNs are as follows, (Kumari et al., 2015), (Wong et al., 2006), and (Ballarin et al., 2013):

**Snooping:** It is a passive kind of attacks that refers to unauthorized access or interception of message contents. It could be prevented by making message contents non-intelligible by using encipherment techniques.

**Traffic Analysis:** Basically, these kinds of attackers notice one communication pattern into the MANET environment.

- Monitor the network traffic: e.g. log files, web pages etc.
- Try to gain useful information from statistical analysis: e.g. who communicates with whom, when, how long, where? Who is interested in what contents etc.?

**Modification:** This is an active kind of attacks. Here, after gaining access over the message, attackers try to modify the information to get own benefit. In this case, also sometimes attackers try to delete or delay the message to harm the system or to get benefit from there.

**Masquerading:** Masquerading or spoofing kind of attack can be mounted in the WMNs when attacker impersonates somebody else. Firstly, one or more legal login request is intercepted by an attacker. Later on, modify this request in such a way so it could pass the authentication test of WMN and get permission to access resources in the network.

**Replaying:** Anyhow attacker obtains the copy of a sent message from the legitimate user to either access the WMN or fraud the legal user by showing itself as a valid service provider. If an attacker gets success, then this attack could be called as replay security attack.

**Repudiation:** It is something different kind of security attack than the previously discussed attacks one. It is performed by one on of the two authorized communicating parties in the WMN either source or destination. In this case, either message sender later deny that he has sent the message or receiver might later deny that he has received it.

**DoS:** It is an active kind of attack and very common in general. It may slow down or totally interrupt the service of a system/network. In this case, attackers can launch several way to achieve this target. They might pump too much fake request in the network that the server crashes because of heavy traffic load. If the attacker gets success to launch this target, then WMN's node becomes unresponsive and nobody will able to connect with it.

Moreover, all kinds of available attacks may be broadly classified into active and passive attacks. Active attacks actively involved in the WMN and try to modify the content of messages or create the disturbance of source and destination authentication or harm the server so it becomes unresponsive. Such attacks are DoS, repudiation, replaying, masquerading, modification, anonymity, fairness that may threaten the integrity of the message, availability of services and authentication of clients. The active attacks are easier to detect then prevent because attackers can launch them in a variety of ways. Whereas the goal of passive attacks is just to obtain information that means attackers doesn't try to modify the content of the message or the server. The disclosing of information may harm the mesh end users client of the message, but the system is not affected. The confidentiality or privacy, authorization, and accounting are threatened by passive attacks. Snooping is traffic analysis fall under passive attacks. These attacks can be prevented by encipherment of the data.

## COUNTERMEASURE AGAINST ATTACKS IN WMNS

The research communities from academia and industry proven that it is very hard to provide fully secure from various attacks in WMNs. Such networks are highly susceptible to attacks due to its openness nature, dynamic topology, and lack of conventional security, unlike their wired counterparts. Nevertheless, the research groups have been put forward to provide several protection techniques against different attacks to reduce the security problem in WMNs. These techniques are fall into three different categories: prevention from intrusion, intrusion detection and intrusion response (Jiacheng et al., 2010). In order to provide secure WMN, this chapter presents several counterparts such as understanding of intrusion prevention scheme, secure routing, intrusion detection approach, etc., majorly focused on intrusion detection system.

### Intrusion Prevention Schemes (IPS)

In order to provide security against various malicious WMN nodes, IPS are considered as the principle line that includes encryption, authentication along with secure routing (Jiacheng et al., 2010). In the literature particularly numerous WMNs key management techniques can be found that entail encryption and authentication. The key management service is mainly responsible to keep track of bindings between keys and WMN nodes. Moreover, also takes care of the establishment of mutual trust and secure communication between clients (Siddique et al., 2007). The already existing key management schemes for wired and wireless can be categorized as follow (Jiacheng et al., 2010):

- **Centralized:** In these schemes, one central trusted party is assigned known as third party or group server, which is mainly liable to generate and distribute the group keys among nodes.
- **Decentralized:** In order to avoid load at a single server, the group management task is distributed to the multiple subgroup servers, unlike the previous one.
- **Distributed Key Management:** This method distributes the keys generation task among one or multiple group members.

Although there are the plethora of key management techniques has proposed by the research community for wired or wireless communication. However, those are not suitable for WMNs environment due to dynamic-nests, openness, multi-hope, heterogeneity of devices nature. Lately, several researchers have developed various key management schemes to suit the WMNs characteristic. Among them, few are tabulated in Table 1.

*Table 1. Various WMNs' key management schemes*

| Authors | Major Contributions | Techniques | Comments |
|---|---|---|---|
| (Zhang et al, 2006) | • Proposed ARSA allows WMN clients to access and roam between multi-domains by eliminating the requirement of bilateral roaming agreements and real-time interaction between potential WMN operators.<br>• Address the user privacy issue by offering distinct alias identities to the user | ARSA is centered on a router-client-client-based identity-based public key cryptosystem for authentication and main agreement (AKA) based on router and client passes. | other severe attacks against WMN need to counterparts like Location Privacy Attack, Bogus-Beacon Flooding Attack, Denial-of-Access Attack, Bandwidth-Exhaustion Attack, etc. |
| (Ding et al., 2011) | Suggested a hierarchical key management system without a certificate.<br>To create identity, each customer must be recorded with an offline trusted party (i.e. offline authentication).<br>Scheme removes the key escrow issue, improves the user's computation and enhances system productivity. | Threshold secret sharing and certificate-less signcryption. | • Node key security, and Forward / backward security is ensured<br>• private key generated by cluster head need to update regularly to improve WMNs security |
| (Wang et al., 2008) | • Presented framework support both one-to-many and many-to-many group communications.<br>• Allow group members to be authenticated from subgroup leaders without revealing their secure information.<br>• Reduced rekeying delay and storage overhead at end clients with minimum communication cost at backbone. | Logical Key Hierarchical, Distributed Threshold Based, Bloom Filter-based /Semi-Anonymity Authentication. | • Security against both insider and outsider attacks.<br>• Study could be extended for different integration of threshold based along with logical key hierarchical techniques. |
| (Fu et al., 2008) | • Support to deal with any size of WMNs & integrated easily with dissimilar networks.<br>• Authorized key & multi signature-based on identity-based cryptosystem & random number are used to create an authorized certificate. Eliminates the problem of broadcasting of own public key<br>• Effectively fight against intermediate attacks. | Virtual Certification Authority (CA), offline CA, identity-based cryptosystem, multi-signature. | • Improve key management in security, expandability, validity, fault tolerance, and usability.<br>• Only developed sophisticated authentication scheme in WMNs future study need to justify claim. |
| (Boudguiga et al., 2012) | • Authentication & key encryption schemes are proposed for IEEE 802.11s networks.<br>• Station authenticates itself to an Authentication Server that delegates the station key generation to Mesh Key Distributors. Here, ID based cryptography is used for shared secret exchange between AS & station and keys' derivation needed to secure the exchanged messages. | ID-based cryptography uses Sakai Kasahara key construction | • authentication scheme provides resistant to the key escrow attack |

# Secure Routing

The routing protocols are the vibrant part in any network that implements network function and finds its efficiency inside WMNs. Owing to special characteristics in WMN such as open medium, changeable topology, nodes' mobility etc. the routing protocols are persistent victims of various attacks trying to reduce their performance. Hence, traditional protocols are not suitable for WMN. WMN's routing protocols must be secure. Therefore, the researchers developed either the mechanism to improve existing MANET's routing protocols like DSR, AODV, DSDV or new security protocols that are appropriate for the WMNs environment. Some typical WMNs routing protocols have been introduced in below Table 2. Moreover, Table 3 shows the different routing phase attacks.

*Table 2. Comparison of secure WMNs routing protocols*

| Authors | Routing Protocols | Suitable for | Kay Idea | Security Technique | Security Against |
|---|---|---|---|---|---|
| (Oliveiro et al., 2008) | AODV-DEX | AODV | Modify hop count values | Reputation-based metric | Gray hole or Sinkhole attacks |
| (Khan et al., 2010, 2012) | SRPM | AODV | Modify AODV's route discovery mechanism | Compute the Unreliability Value of the neighbours & two-hop passive acknowledgment scheme | Robust against Black hole, Grey hole, Wormhole, Fairness reduction, Jellyfish, & Node isolation. |
| (Wu et al., 2012) | Onion Ring (Novel Communication Protocol) | Onion routing algorithm | Hiding routing information& Protect node privacy by using both cryptography and redundancy. | "Onion", i.e., layered encryption, & padding Techniques. | Inside& aggressive attackers, global adversary, traffic analysis, and eavesdropping |
| (Hui et al., 2012) | Privacy-Aware Secure Hybrid Wireless Mesh Protocol (PA-SHWMP) | SHWMP | Security classification and reputation computation, packet authentication, Routing confidentiality | Dynamic Reputation Mechanism based on subject logic and uncertainty with the multi-level security technology | Assure authenticity, integrity, routing packets' secrecy and internal attacks due to compromised WMN routers, however, it is vulnerable to attacks launched by internal legitimate mesh routers. |
| (Li et al., 2011) | Security Enhanced AODV (SEAODV) | AODV or (enhance on demand part of HWMP) | Authenticate unicast and broadcast Routing messages | Blom's key pre-distribution scheme to compute the pairwise transient key | Identified routing attacks, such as Route REQuest flooding, RREP routing loops, route redirection, and formation of routing loops, |
| (Islam et al., 2009) | Secure HWMP (SHWMP), | HWMP | Modification of the HWMP, cryptographic extensions to provide authenticity and integrity messages | Merkle Tree Concept to authenticate mutable information, symmetric key encryption to protect mutable fields in the routing information elements. | Authenticity & integrity of HWMP routing messages, Prevent unauthorized manipulation of mutable fields. However, vulnerable to attacks launched by internal legitimate mesh routers. |
| (Ben-Othman, et al., 2011) | Enhanced security level of HWMP | HWMP | Modify trust management for internal nodes & digital signature for external nodes. | Identity Based Cryptograph | Authenticity of public keys and ensures integrity of control message in HWMP |

*Table 3. Security attacks on different routing phases*

| Routing Protocol Phases | Description of Security Attacks |
|---|---|
| Route Discovery Phases | • Routing table overflow<br>• Routing cache positioning |
| Route Maintenance Phases | • False route control message |
| Packets Forwarding Phases | • Data dropping<br>• Wormhole attack |
| Advanced Attacks | • Blackhole/sinkhole attack<br>• Byzantine attack<br>• Rushing attack<br>• Resource consumption attack<br>• Location disclosure attack |

# Intrusion Detection Systems (IDSs)

WMNs could not be effectively and efficiently protected with the help of cryptography and protection software. Therefore, IDSs also need to implement to provide the second line of defense Wang et al. (2009). IDSs are used to inform the clients about various possible attacks, stop it or recover from loss both in wired or wireless media (Deb et al., 2011), and (Chen et al., 2009). The different functions of IDS are as follow:

- **Event Monitoring:** IDS must be able to monitor different type of events and maintain the past information related to these events.
- **Analysis Engine:** there must be provisioned to provide analysis engine that processes the collected data to spot unusual or malicious activities.
- **Response:** IDS must be able to alert system administrators (I. e. generate a response).

Traditionally, IDS are not suitable to use in WMNs. (Zhang et al., 2003). Here authors develop specific design of intrusion detection and response mechanisms. (Marti et al., 2000) proposed two schemes: Watchdog and Path-rater to improve efficiency in the network in presence of malicious nodes that agree for packets' forwarding but unable to do it. In WMNs environment, cooperation among nodes are highly needed to support various basic network's functionalities. To impose co-operation among network's clients used mechanisms are token-based, credit-based, and reputation-based, etc. IDS tries to gather various activities information from all available clients. And, then with the help of this information, tries to identify any malicious activity that violating the network security rule. These unusual activities are considered as attacks and accordingly inform security administrator. Moreover, it initiates a proper response to malicious activity. On the basis of architecture, the IDSs can be also classified into three different categories:

- **Stand-Alone IDSs:** Each node deploys IDS mechanism independently to determine intrusions.
- **Distributed and Cooperative IDSs:** Every client takes part in the intrusion detection process and response by having an IDS agent running on them. An IDS agent is responsible for the various tasks like to detect and collect local events, identify possible intrusions, initiate response independently (Zhang et al., 2003).
- **Hierarchical IDSs:** The cluster heads act as the central controller to provide IDS functionalities on its child nodes.

Distributed IDS and Hierarchical IDS are more appropriate for WMNs because to have separate IDS on each mobile client is not feasible. Again, due to WMNs characteristics, IDSs of wired and another wireless media is not applicable in this network. The number of works has been found in the literature in order to detect intrusion in the WMNs that are summarized in the Table 4:

## DIGITAL SIGNATURE-BASED IDS

Although, in the literature, various IDS approaches have been found in order to detect and prevent the intrusions and few of them are tabulated in the table. 4 that are appropriate for the characteristics of WMNs. However, these various intrusion detection methods are not able to provide complete secure WMNs like reputation and self-organizing based IDS (Bankovic et al., 2010); not fast enough to prevent neighbor nodes from being affected, Open LIDS (Hugelshofer et al., 2009); not able to distinguish between RTP stream & UDP DoS flood and not effective for new connections, etc. Moreover, the various other issues of existing IDS are pointed out in the literature such as false identity problem, not suitable to provide trustworthy routes, not able to detect real time attacks, false route control message, participating of nodes those having limited transmission power, etc. (Ping et al., 2008), (Martignon et al., 2009), and (Glass et al., 2008).

Nowadays, due to the huge proliferation of mobile users in wireless networking, one of the most identification and authentication mechanisms to provide security in the digital world is the use of digital signature. It is becoming popular because of its various inherent features such as forging of a digital signature is generally computationally hard, identification and verification of the digital signature is relatively easy, problem of both forgery and DoS by receiver and sender can be easily eliminated, in case of any dispute, it is able to verify by any other third parties, etc. In the literature, there is number of most important digital signatures that are more frequently used such as RSA, ElGamal, Rabin algorithm etc., have been presented. It is expected that with the help of any of the digital signature scheme, one can easily provide the secure acknowledgment and authentication in WMN's client, router, server, and packet level. The flaws of WMN because of existing IDS may be removed with the help of digital signature-based IDS. The digital signature-based IDS may ensure security at two layers. In the first layer, extra reserved bits are used to maintain sequence number; this is done for both packet and ACK transmission. Further, the second layer provides the double safeguard to forward packets, by putting a digital signature.

## Table 4. Summary of different IDS for WMNs

| IDS Reference | Underlying Protocol | Addressed Attacks | Architecture | Remarks |
|---|---|---|---|---|
| (Bansal et al., 2011). Detecting MAC layer greedy misbehaviour | Not specified | Oversized NAV attack, reduced back off attack, Hybrid and Fast switching attacks | Distributed System | • Switching or detection of smart attacks.<br>• IDS is implemented on Mesh points, more cost effective.<br>• Suitable: for MANET and WMN<br>• Enhanced to detect new selfish MAC misbehaviors. |
| (Bankovic et al., 2010). Reputation systems & self-organizing maps. | Not specified | Resource utilization & Routing misbehaviour attacks | Distributed agent-based Systems | • Confidentiality and integrity cannot be preserved for any node.<br>• Reputation system identifies the attacked node immediately.<br>• Not fast enough to prevent neighbor nodes from being affected. |
| (Zhangan et al., 2008). Reputation-based IDS (RADAR) | DSR | Malicious nodes, DOS Attack, Routing Loop Attack. | Distributed System | • Routing route with higher false alarms.<br>• Resilient to malicious collectives for subverting reputations.<br>• High latency to detect DoS attacks. |
| (Glass et al., 2009). MAC-layer ID mechanism | Not specified | Detecting man-in-the-middle, wormhole attacks, exploits positive acknowledgment property | Distributed & Centrally Control System | • High detection rate,<br>• No false positives,<br>• Small computational and communication overhead<br>• Small loss of bandwidth. |
| (Zaidi et al., 2009). Principal Component Analysis | Not specified | DoS, Port Scan Attacks etc. | Distributed System | • Reduced number of false alarms<br>• Node outages are not detected, method looks for spurious traffic generation.<br>• Not consistent due to unrealistic assumptions on network traffic. |
| (Hugelshofer et al., 2009). Open LightweightIDS (OpenLIDS) | Not specified | Resource starvation attacks, spam email distribution, mass mailing of internet worms, IP spoofing. | Distributed System | • Produced no false positives with traffic traces.<br>• OpenLIDS not able to distinguish between RTP stream & UDP DoS flood.<br>• Not effective for new connections.<br>• Arbitrarily adjust timeout values is not suitable. |
| (Khan et al., 2010). Cooperative & cross layer IDS | Not specified | Multilayer security attacks | Distributed System | • Client collectively monitor the neighbors, if intrusion found, report is sent to mesh router, and action taken based on severity level.<br>• Mesh routers also collectively monitor the neighbors, if a malicious mesh router is found, information is broadcast to all the neighbors.<br>• However, Detection & response of cross layer parameters exchange in WMN is still not addressed that could be avoided by intrusion signatures |
| (Shila et al., 2008). Trust based approach: Counter-Threshold & Query Based | AODV Also applied for DSR, DSDV | Selective forwarding attack or Gray hole attacks. | Distributed System | • Efficient against presence of selective forwarding attackers.<br>• Overhead increases with number of attackers.<br>• If detection threshold<throughput of a path, then attacks will not be detected and network throughput will suffer.<br>• On the contrary, if detection threshold> throughput of the path, the throughput would suffer even if there is no attacker. |
| (Ping et al., 2008). Timed automata-based distributed IDS | DSR | Unknown intrusion with false alarms. | Distributed MANETs | • Network monitors the traced data on each node, and audit every forwarding packet by automata<br>• Detect real-time attacks without signatures of intrusion. |
| (Martignon et al., 2009). Trust & reputation management IDS | AODV | Selfish behaviour of mesh routers that participate in community network, bad-mouthing attack | Heterogeneous mesh systems | • Offers high detection accuracy, even high %of network nodes provide false trust values.<br>• However, scheme is not trusty for selfish behaviors in multi-channel environments.<br>• Availability of more alternative paths, not suitable to provide trustworthy paths. |
| (Yang et al., 2010). Proxy-based IDS | Not specified | Active attacks: forging, retransmitting, distorting information | distributed proxy servers | • IDS proxy runs independently and detects the activities of inner nodes<br>• Cross domain intrusion detection is cooperatively done through gateway nodes |

## FUTURE RESEARCH DIRECTIONS

In order to implement the concept of digital signature-based IDS at various levels such as client, server, router, etc. in WMNs for securing acknowledgment and authentication, a proper selection of digital signature scheme is needed. The various digital signature schemes are available in the literature such as RSA, ElGamal, Rabin, Digital Signature Standard (DSS), Hash Function, Elliptic Curve Digital Signature Algorithm (ECDSA), Boneh Lynn Shacham (BLS) signature, etc. The users may choose any one of the signature schemes as per their own constraint and requirement like available bandwidth, memory, power, etc. because each and every scheme having their own strength and weakness.

## CONCLUSION

The wireless multimedia network has recently gained lots of significance in view of its commitment to provide better services in the next generation of wireless networking, multimedia applications and uses. WMNs due to its potential applications in emergency and disaster zone, peer to peer communication, health and medical field, hiding information, building automation, transportation system, broadband home, metropolitan area networking, etc., has become a vibrant area among the research communities. WMN is self-configuring, self-motivated, arbitrary, and multi-hop network that is composed of bandwidth constrained wireless links without an aid of centrally controlled routers/servers. Owing to these characteristics such networks got number of advantages like reliable service coverage, robustness against failure, easy network deployment and maintenance, low cost etc. However, because of its open nature, implementation of security aspect becomes one of the paramount concerns WMN.

The conclusion is supported with the fact that the various security problems in WMNs can be achieved up to a good level. The security issues that evolved with the misbehaving and selfishness nature of nodes within WMNs, as nodes in the same network have open characteristics nature with limited battery life. Owing to misbehaving and selfishness nature of nodes, it becomes easy for unauthorized users to attack the network and access the valid information. So it is proposed that the digital based signature IDS might solve these challenges by identifying the malicious attacks and then preventing them from becoming the malicious nodes (adversary nodes). Thus, the information exchanging among the users using WMNs will become secure with basic security issues like confidentiality, integrity, authorization etc. and can vanquish problems of reputation, receiver collision, false identity problems etc.

The major findings of this chapter after literature work are illustrated as follows:

- The major security issues are still open research challenging problem in Wireless Mesh Networks. To solve these problems, deep learning techniques and customized IDS with new approaches and digital signature-based credentials are used to provide a fundamental solution for it.
- Once digital signature-based IDS will successfully implement in the WMNs then the clients would able to achieve satisfied communication within the network.
- The digital signature-based IDS can achieve these issues by providing the security at all levels of the network viz. client, router, server and packet level. The digital signature-based IDS in this digital world are supposed to provide identification and authentication among the huge proliferation of users accessing wireless multimedia through WMNs.
- New exploration of the frameworks and algorithms are designed to solve these security issues.

# REFERENCES

Akyildiz, I., & Wang, X. (2009). *Wireless Mesh Networks (Advanced Texts in Communications and Networking)*. Chichester, UK: John Wiley & Sons.

Akyildiz, I. F., Wang, X., & Wang, W. (2005). Wireless Mesh Networks: A survey. *Computer Networks*, *47*(4), 445–487. doi:10.1016/j.comnet.2004.12.001

Asherson, S. & Hutchison, A. (n.d.). Secure routing in wireless mesh networks, *University of Cape Town*. Retrieved from http://pubs.cs.uct.ac.za/archive/00000318/01/SATNAC2006WIP.pdf

Ballarini, P., Mokdad, L., & Monnet, Q. (2013). Modeling tools for detecting DoS attacks in WSNs. *Security and Communication Networks*, *6*(4), 420–436. doi:10.1002ec.630

Bankovic, Z., Fraga, D., Manuel, M. J., Carlos, V. J., Malagon, P., Araujo, A., & Nieto-Taladriz, O. (2010). Improving security in WMNs with reputation systems and self-organizing maps. *Journal of Network and Computer Applications*, *34*(2), 455–463. doi:10.1016/j.jnca.2010.03.023

Bansal, D., Sofat, S., Pathak, P., & Bhoot, S. (2011). Detecting MAC misbehaviour switching attacks in Wireless Mesh Networks. *International Journal of Computers and Applications*, *26*(5), 55–62. doi:10.5120/3102-4261

Ben-Othman, J., & Benitez, Y. I. S. (2011). IBC-HWMP: A novel secure identity-based cryptography-based scheme for Hybrid Wireless Mesh Protocol for IEEE 802.11s. *Concurrency and Computation, 25*(5), 686–700. doi:10.1002/cpe.1813

Ben-Othman, J., & Benitez, Y. I. S. On securing HWMP using IBC, *IEEE International Conference on Communications, Kyoto, Japan*, 1–5. doi:10.1109/icc.2011.5962921

Boudguiga, A., & Laurent, M. (2012). An authentications scheme for IEEE802.11s mesh networks relying on Sakai-Kasahara ID-Based Cryptographic Algorithms, *International Conference on Communications and Networking, Niagara Falls*, 256–263. doi:10.1109/comnet.2012.6217728

Chen, T., Kuo, G.-S., Li, Z.-P., & Zhu, G. M. (2009). Intrusion detection in Wireless Mesh Networks Security, Wireless Mesh Networks. Boca Raton, FL: Auerbach Publications.

Choudhary, G., Sharma, V., You, I., Yim, K., Chen, I. R., & Cho, J. H. (2018). Intrusion detection systems for networked unmanned aerial vehicles: A survey. *Proceedings of the 14th International Wireless Communications & Mobile Computing Conference, Limassol, Cyprus*, 560-565.

Deb, N., Chakraborty, M., & Chaki, N. (2011). A state-of-the-art survey on Ids for mobile ad-hoc networks and wireless mesh networks. *Communications in Computer and Information Science, 203*, 169–179. doi:10.1007/978-3-642-24037-9_17

Durahim, A. O., & Savaş, E. (2010). A-MAKE: An efficient, anonymous and accountable authentication framework for WMNs. In *Proceedings of 5th International Conference on Internet Monitoring and Protection, Barcelona, Spain*, 54-59. 10.1109/ICIMP.2010.16

Egners, A., & Meyer, U. (2010). Wireless Mesh Network Security: State of Affairs. In *Proceedings of 6th IEEE Workshop on Security in Communication Networks, Denver, Colorado*, 997-1004, 10.1109/LCN.2010.5735848

Feng, Y., Fan, M. Y., & Liu, C. P. (2008). A new privacy-enhanced authentication scheme for Wireless Mesh Networks. *School of Computer Science and Technology. University of Electronic Science and Technology of China, 265-269.* doi:10.1109/ICACIA.2008.4770020

Fu, Y., He, J., Luan, L., Wang, R., & Li, G. (2008). A zone-based distributed key management scheme for Wireless Mesh Networks. In *Proceedings of 32nd Annual IEEE International Computer Software and Applications Conference, Turku, Finland*, 68–71. 10.1109/COMPSAC.2008.131

Fu, Y., He, J., Wang, R., & Li, G. (2008). Mutual authentication in Wireless Mesh Networks. In *Proceedings of IEEE International Conference on Communications*, Beijing, China, 1690–1694. doi:10.1109/icc.2008.326

Gao, L., Chang, E., Parvin, S., Han, S., & Dillon, T. (2010). A secure key management model for Wireless Mesh Networks. In *Proceedings of 24th IEEE International Conference on Advanced Information Networking and Applications, Perth, WA*, 655–660. 10.1109/AINA.2010.110

Glass, S., Portmann, M., & Muthukkumarasamy, V. (2008). Securing Wireless Mesh Networking. *IEEE Internet Computing*, *12*(4), 30–36. doi:10.1109/MIC.2008.85

Glass, S. M., Muthukkumurasamy, V., & Portmann, M. (2009). Detecting man-in-the-middle and wormhole attacks in Wireless Mesh Networks. In *Proceedings of International Conference on Advanced Information Networking and Applications, Bradford, UK*, 530–538. 10.1109/AINA.2009.131

Hamid, Z., & Khan, S. A. (2006). An Augmented Security Protocol for Wireless MAN Mesh Networks. In *Proceedings of International Symposium on Communications and Information Technologies*, 861-865. doi:10.1109/iscit.2006.339859

Hector, M., Guha, L. C., & Lu, R. K. (2011). A secure service-oriented routing algorithm for heterogeneous wireless mesh networks. In *Proceedings of IEEE Global Telecommunications Conference*, 1-5. doi: 10.1109/GLOCOM.2011.6134439

Hugelshofer, F., Smith, P., Hutchison, D., & Race, N. J. P. (2009). OpenLIDS: A lightweight intrusion detection system for wireless mesh networks. In *Proceedings of 15th Annual International Conference on Mobile Computing and Networking, China*, 309-320. 10.1145/1614320.1614355

Islam, S., Hamid, A., & Hong, C. S. (2009). *SHWMP: a secure hybrid wireless mesh protocol for IEEE 802.11s wireless mesh networks. Transactions on Computational Science, 5730* (pp. 95–114). Berlin, Germany: Springer-Verlag.

Jiacheng, H., Ning, L., Ping, Y., Futai, Z., & Qiang, Z. (2010). Securing Wireless Mesh Network with mobile firewall. In *Proceedings of IEEE International Conference on Wireless Communications and Signal Processing*, Suzhou, China, 1–6. 10.1109/WCSP.2010.5633566

Khan, S., Leo, K. K., & Din, Z. U. (2010). Framework for intrusion detection in IEEE 802.11 Wireless Mesh Networks. *The International Arab Journal of Information Technology, 7*(4), 435–440.

Khan, S., Loo, K. K., Mast, N., & Naeem, T. (2010). SRPM: Secure routing protocol for IEEE802.11 infrastructure based wireless mesh networks. *Journal of Network and Systems Management, 18*(2), 190–209. doi:10.100710922-009-9143-3

Khan, S., Nabil, A. A., & Loo, K. K. (2012). Secure route selection in Wireless Mesh Networks. *Computer Networks, 56*(2), 491–503. doi:10.1016/j.comnet.2011.07.005

Kumari, S., Khan, M. K., & Atiquzzaman, M. (2015). User authentication schemes for wireless sensor networks: A review. *Ad Hoc Networks, 27*, 159–194. doi:10.1016/j.adhoc.2014.11.018

Li, C., Wang, Z., & Yang, C. (2011). Secure routing for Wireless Mesh Networks. *International Journal of Network Security, 13*(2), 109–120.

Li, Z. & Guang, G. (2008). *A survey on security in wireless sensor networks*. Retrieved from HTTP://CACR.UWATERLOO.CA/TECHREPORTS/2008/CACR2008-20.PDF

Lin, H., Ma, J., Hu, J., & Yang, K. (2012). PA-SHWMP: A privacy aware secure hybrid wireless mesh protocol for IEEE 802.11s wireless mesh networks. *EURASIP Journal on Wireless Communications and Networking, 69*, 1–16. doi:10.1186/1687-1499-2012-69

Malik, R., Mittal, M., Batra, I., & Kiran, C. (2011). Wireless Mesh Networks (WMN). *International Journal of Computers and Applications, 1*(23), 68–76. doi:10.5120/533-697

Marti, S., Giuli, T., Lai, K., & Baker, M. (2000). Mitigating routing misbehaviour in mobile ad hoc networks. In *Proceedings of Sixth Annual International Conference on Mobile Computing and Networking, Boston*, MA, 255-265. doi:1910.3459550.1145/345

Martignon, F., Paris, S., & Capone, A. (2009). A framework for detecting selfish misbehaviour in Wireless Mesh Community Networks. In *Proceedings of 5th ACM International Symposium on QoS and Security for Wireless and Mobile Networks, Tenerife, Canary Islands, Spain*, 65-72. doi:10.1145/1641944.1641958

Naveed, A., Kanhere, S. S., & Jha, S. K. (2009). Attacks and security mechanisms security in wireless mesh networks. Boca Raton, FL: Auerbach Publications.

Oliviero, F., & Romano, S. P. (2008). A reputation-based metric for secure routing in wireless mesh networks. In *Proceedings of IEEE Global Telecommunications Conference, New Orleans, LA*, 1–5. 10.1109/GLOCOM.2008.ECP.374

Ping, Y., Xinghao, J., Yue, W., & Ning, L. (2008). Distributed intrusion detection for mobile adhoc networks. *Elsevier Journal of System Engineering and Electronics, 19*(4), 851-859. doi: (08)60163-2 doi:10.1016/S1004-4132

Reed, M. G., Syverson, P. F., & Goldschlag, D. M. (1998). Anonymous connections and onion routing. *IEEE Journal on Selected Areas in Communications, 16*(4), 482–494. doi:10.1109/49.668972

Ren, K., Yu, S., Lou, W., & Zhang, Y. (2010). PEACE: A novel privacy-enhanced yet accountable security framework for metropolitan wireless mesh networks. *IEEE Transactions on Parallel and Distributed Systems, 21*(2), 203–215. doi:10.1109/TPDS.2009.59

Salem, N. B., & Hubaux, J. P. (2006). Securing Wireless Mesh Networks. *IEEE Wireless Communications, 13*(2), 50–55. doi:10.1109/MWC.2006.1632480

Seyedzadegan, M., Othman, M., Ali, B. M., & Subramaniam, S. (2011). Wireless Mesh Networks: WMN Overview, WMN Architecture, *International Conference on Communication Engineering and Networks*, Singapore, 12-18.

Sgora, A., Vergados, D. D., & Chatzimisios, P. (2016). A survey on security and privacy issues wireless mesh networks, security and communication networks. *Wiley Online Library, 9*(13), 1877–1889. doi:10.1002ec.846

Shila, D. M., & Anjali, T. (2008). Defending selective forwarding attacks in WMNs. In *Proceedings of IEEE International Conference on Electro/Information Technology, USA*, 96-101. doi:10.1109/eit.2008.4554274

Siddiqui, M. S., & Hong, C. S. (2007). Security issues in Wireless Mesh Networks. In *Proceedings of International Conference on Multimedia and Ubiquitous Engineering, Seoul, Korea*, 717–722, doi:10.1109/mue.2007.187

Stajano, F., & Anderson, R. (2002). Resurrecting duckling: Security issues for ubiquitous computing. *Supplement to Computer, 35*(4), 22–26. doi:10.1109/mc.2002.1012427

Stallings, W. (2006). Network security essentials (3rd ed.). Upper Saddle River, NJ: Prentice Hall.

Tong, Y. P., Liu, N., & Wu, Y. (2009). Security in wireless mesh networks: Challenges and solutions. In *Proceedings of Sixth International Conference on Information Technology: New Generations*, Las Vegas, NV, 423-428. doi:10.1109/ITNG.2009.20

Trong, M. H., Dinh, V. L., & Kim, N. Q. (2013). A study on routing performance of 802.11 based wireless mesh networks under serious attacks. In *Proceedings of International Conference on Computing, Management and Telecommunications*, 295-297. doi:10.1109/ComManTel.2013.6482408

Vural, S., Wei, D., & Moessner, K. (2013). Survey of experimental evaluation studies for wireless mesh network deployments in urban areas towards ubiquitous internet. *IEEE Communications Surveys and Tutorials*, *15*(1), 223–239. doi:10.1109/SURV.2012.021312.00018

Wang, X., Patil, A., & Wang, W. (2006). VoIP over wireless mesh networks Challenges and approaches. In *Proceedings of 2nd Annual International Workshop on Wireless Internet*, 1-9. doi:10.1145/1234161.1234167

Wang, X., Wong, J., & Zhang, W. (2008). A heterogeneity-aware framework for group key-management in wireless mesh networks. In *Proceedings of 4th International Conference on Security and Privacy in Communication Networks*, Instabul, Turkey. 10.1145/1460877.1460918

Wang, X., Wong, J. S., Stanley, F., & Basu, S. (2009). Cross-layer based anomaly detection in Wireless Mesh Networks. In *Proceedings of 9th Annual International Symposium on Applications and the Internet*, Bellevue, WA, 9–15. 10.1109/SAINT.2009.11

Wong, K. H. M., Zheng, Y., Cao, J., & Wang, S. (2006). A dynamic user authentication scheme for wireless sensor networks. In *Proceedings of IEEE International Conference Sensor Networks, Ubiquitous, Trustworthy Computing*, 244–251. IEEE Computer Society. 10.1109/SUTC.2006.1636182

Wu, X., & Li, N. (n.d.). Achieving privacy in mesh networks. *4th ACM Workshop on Security of Ad Hoc and Sensor Networks*, Alexandria, VA, 13–22. doi:10.1145/1180345.1180348

Yang, Y., Zeng, P., Yang, X., & Huang, Y. (2010). Efficient intrusion detection system model in Wireless Mesh Network; Networks Security Wireless. In *Proceedings of 2nd International Conference on Communications and Trusted Computing*, Wuhan, China, 393–395. doi:10.1109/NSWCTC.2010.226

Yi, D., Xu, G., & Minqing, Z. (2011). The research on certificate less hierarchical key management in Wireless Mesh Network. In *Proceedings of 3rd IEEE International Conference Communication Software and Networks*, Xian, China, 504-507. doi:10.1109/iccsn.2011.6013643

Zaidi, Z. R., Hakami, S., Landfeldt, B., & Moors, T. (2009). Detection and identification of anomalies in wireless mesh networks using Principal Component Analysis (PCA). *Journal of Interconnection Networks, 10*(04), 517–534. doi:10.1142/S0219265909002698

Zhang, Y., & Fang, Y. (2006). ARSA: An attack resilient security architecture for multi-hop wireless mesh network. *IEEE Journal on Selected Areas in Communications, 24*(10), 1916–1928. doi:10.1109/JSAC.2006.877223

Zhang, Y., Lee, W., & Huang, Y. (2003). Intrusion detection techniques for mobile wireless networks, *ACM/Kluwer Wireless Networks Journal, 9*(5), 1-16.

Zhang, Y., Luo, J., & Hu, H. (2006). Wireless mesh networking: Architectures, protocols and standards. Boca Raton, FL: Taylor & Francis Group.

Zhangm, Z., Nait-Abdesselam, F., Ho, P.-H., & Lin, X. (2008). *RADAR: A reputation-based scheme for detecting anomalous nodes in wireless mesh networks.* In Proceedings of *IEEE Wireless Communications & Networking Conference,* Las Vegas, NV. (pp. 2621–2626). doi:10.1109/wcnc.2008.460

# Chapter 2
# Industrial Automation Using Internet of Things

**Samyak Jain**
*National Institute of Technology Karnataka, Surathkal, India*

**K. Chandrasekaran**
*National Institute of Technology Karnataka, Surathkal, India*

## ABSTRACT

*This chapter presents a comprehensive view of Industrial Automation using internet of things (IIoT). Advanced Industries are ushering in a new age of physical production backed by the information-based economy. The term Industrie 4.0 refers to the 4th paradigm shift in production, in which intelligent manufacturing technology is interconnected with physical machines. IIoT is basically a convergence of industrial systems with advanced, near-real-time computing and analytics, powered by low cost and low power sensing devices leveraging global internet connectivity. The key benefits of Industrial IoT systems are a) improved operational efficiency and productivity b) reduced maintenance costs c) improved asset utilization, monitoring and maintenance d) development of new business models e) product innovation and f) enhanced safety. Key parameters that impact Industrial Automation are a) Security b) Data Integrity c) Interoperability d) Latency e) Scalability, Reliability, and Availability f) Fault tolerance and Safety, and g) Maintainability, Serviceability, and Programmability.*

DOI: 10.4018/978-1-7998-0373-7.ch002

## BACKGROUND

### Internet of Things

The term Internet of things (IoT) was coined in 1999 by Kevin Ashton of P&G (Procter & Gamble). IoT basically is an interconnection of objects such as appliances, devices, vehicles and other items, broadly termed as "things". It comprises of devices such as sensors and actuators, hardware and firmware electronics, system and application software, and finally the connectivity which "enables" objects to link together and interchange data. Each "thing" has an address to uniquely identify the object and an ability to connect and operate with existing internet infrastructure. As per the estimates, IoT will span approximately 30 billion objects by 2020. The global market value of IoT is estimated at $7.1 trillion growing at a healthy compounded annual growth rate. IoT is blurring the lines between the physical and digital world. This meshed world is popularly known as "phygital" world. IoT enables "things" to be sensed, configured, monitored and operated (controlled) remotely leveraging the existing internet infrastructure. The key benefits of IoT are improved operational efficiency and productivity through automation, energy conservation, better precision and accuracy, improved safety and security and economic benefits like increased revenue and reduced expenses with reduced human intervention. Further, IoT is slated to improve end-user experience, engagement and satisfaction.

### Industrial Internet of Things

The term Industrial Internet of Things (IIoT) was defined by General Electric [GE] in 2012. Industrial Internet is basically embedding of devices such as sensors, actuators and other similar instrumentation in machines to create a world of Smart Machines. IIoT, in a true sense provides a platform to converge global industrial systems using low-cost/low-power sensing devices that generate "Big Data" [High volume, High Velocity and High Variety] by adopting advanced analytics and computing. The platform is an interconnected mesh comprising of Machine-to-Machine [M2M] and People-to-Machine [P2M] and Machine-to-People [M2P]. IIoT meshes the "Industrial" world with "Digital" world and has the potential to transform and automate global Industries. The paradigm shift being "industrial data" as the source of competitive advantage which can be processed "anywhere" in a "hyper-connected" world. IIoT provides capability to manufacturing organizations to collect, aggregate and analyze large amounts of sensitive machine data in near real time mode to configure, monitor, manage and maintain machine performance and availability. Further, aggregated machine performance can improve efficiency of connected Factories/ Plants/ Assembly lines. Finally, the data collected could itself be smart and route itself to right users/ user community for real-time decision making. Figure 1 depicts the Components of an Industrial IoT Solution.

IIoT has tremendous potential, an estimate suggests that IIoT Solutions could add $32.3 trillion to global GDP. This represents a 46% of global economy today. The key elements of Industrial IIoT Solutions (Peter & Marco, 2012) are:

- **Intelligent/ Smart Machines**: Connected machines, fleets (e.g. airplanes, vehicles), facilities (buildings) and networks with sensors, actuators and application software.
- **Advanced Analytics:** Combination of descriptive, predictive and prescriptive algorithms to generate real time insights for improved decision making and high end automation. Advanced Analytics requires deep understanding of both domain and attribute/ feature data.
- **People:** Connecting people anywhere, anytime with any-device using any-path/any network. This is also termed as Internet of People.

Figure 2 depicts the typical data flow from Connected Machines in the form of an Industrial IoT Data loop.

Gartner predicts that by 2021 a million IoT devices will probably be installed every hour. Further, Gartner has also forecasted that there shall be 35 Billion connected "things" on the internet by 2020 and that a large proportion of these devices (47%) will be intelligent devices.

*Figure 1. Components of industrial IoT solutions*

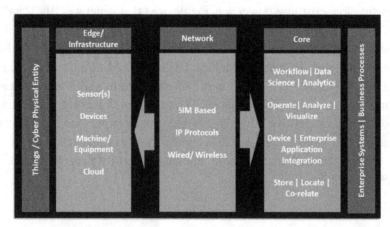

*Figure 2. Industrial IoT data loop*

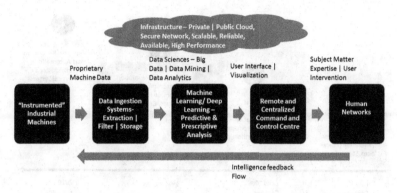

# EVOLVING INDUSTRIAL LANDSCAPE

Today, the world is witnessing an unprecedented era of disruptive innovation with the advent of Industrial Internet of Things (IIoT). The term Industrie 4.0 or Industrial Internet of things is essentially the fourth paradigm shift in manufacturing industry, in which intelligence is built into manufacturing ushering in an era of Smart Manufacturing. Industrie 4.0 is really a "digital" revolution that uses digital technologies and automation to create a future state vision for the manufacturing industry. A vision that is based on the concept of Smart connected factories manufacturing Smart products (Maqbool, Xiaotong, & Wanchun, 2017) for a better world.

The first three paradigms were mechanization (steam engine), electrification (conveyor belt) and computerization (PLC/ Numerically controlled machines). Figure 3 below depicts the historical evolution of Industrial Revolution (Brenna, Monika, & Mark, 2016).

As the history of Industrial Revolution shows, Manufacturing has been evolving ever since its inception in 1850s when the products were mainly hand-crafted. In early to mid- twentieth century [1900-1950] the drive was to build standard products in mass volumes [e.g. Ford Cars] with an aim to optimize costs. As prosperity increased, people started to demand customized products. Off-late, especially in the last 2 decades customers have become very demanding and there is a need to adapt to the frequently changing customer demands. Figure 4 demonstrates the evolution of Manufacturing Products (Daryll et al., 2015).

*Figure 3. History of industry revolution*

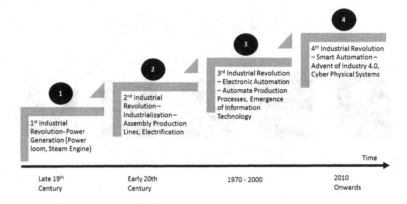

*Figure 4. Product evolution*

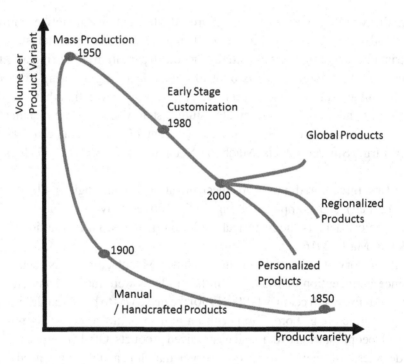

## Evolution of Customer Needs

Since 2000, topics like globalization/ regionalization, personalization, regulatory compliances, eco-friendliness have gathered momentum. While companies want to develop Global Products at mass volumes [one size fits all] or at best develop customized regional products reflecting the demographics of a region, the customer demand really is to develop "personalized" products in much smaller batch sizes, eventually a batch-size of 1 – this represents a challenge of Product Variety vs Volume to the manufacturing organizations.

Modern Factories, therefore, need to manufacture customized products (mass customization/ personalization), in small lots/ batch sizes (variable volumes), with high variability of product types (new designs) and with frequent changes in product portfolio. This essentially means that the same machine/ assembly line and plant needs to be dynamically configured/ programmed near real-time to produce products in line with the Customer demand, in variable quantities and at optimized costs. This truly represents, the dilemma that the Manufacturers are facing today.

Secondly, according to the Smile Curve Manufacturing Theory (Daryll et al., 2015) processes at the end of production value chain i.e. Product Research and Design and Product Marketing and Support functions drive higher financial value as compared to the manufacturing function that supposedly creates least value. Refer to Figure 5 which depicts the Manufacturing Curve (Adapted from "The Stan Shih Smile Curve") – as can be seen, the real act of producing goods creates least financial value.

The Manufacturing Organizations, therefore, have to be both flexible and extremely cost efficient. Table 1 summarizes the emerging manufacturing flexibility needs (Daryll et al., 2015).

*Figure 5. Manufacturing s(mile) curve*

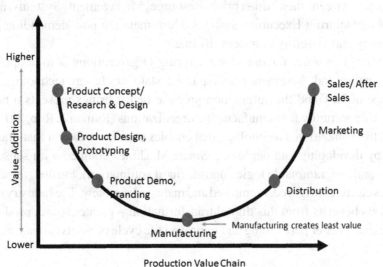

*Table 1. Manufacturing flexibility*

| Flexibility Need | Description |
|---|---|
| Process | Number of Components/Parts that can be manufactured without a need for major set-up change |
| Delivery | Ability of system to respond to changing delivery requests |
| Market | Ability of system to respond to changing market needs [Macro trends] |
| New Design | Lead time to design and introduce new products [Speed-to-Market] |
| Product Variant | Lead time to create a product variant and/or substitute new parts |
| Volume | Ability of system to respond to changing need for product volumes |
| Automation | Extend of Flexibility embedded in manufacturing automation |

In a nutshell, global manufactures today are expected to balance and manage the Customer needs in terms of designing new products, to designing innovative new features/ variants for existing products, to responding to new Market segments, to changing product volumes at will – almost all key assumptions and static variables of yester years have now become dynamic variables. With the average life of a manufacturing organization steadily decreasing to approximately 20 years, there is a significant pressure and an opportunity to create and adopt *dynamic, self-adapting/ learning and autonomous Industrial systems.*

Further, with supply chains becoming more complex and dynamic, customers today expect better co-operation and greater transparency from their manufacturing suppliers, in terms of order tendering, preparation and fulfillment and an automated integration between their Enterprise Resource Management Systems and the actual Manufacturing Execution Systems. Customers are now demanding greater transparency and visibility in order fulfillment.

The need, therefore, for the Manufacturing Organizations is to adopt Digital Transformation and Automation using latest state of the art technologies. The emergence of IIoT and the automation promise that it brings presents a business disruption opportunity for manufacturing organizations (Raman & Rob, 2017). IIoT provides the foundational technology that enables transformation of manufacturing process by developing and deploying Smart Machines coupled with sensors and actuators and programmable logic controls (hard automation). Further, these Smart Machines can be configured, monitored and managed remotely. The Industry expects tremendous benefits from this digital transformation – greater speed, productivity and efficiency, better performing and improved life cycle of assets (plants/ assembly lines/ machines), sturdier economic growth, and better/ safer jobs resulting in improved living standards across the globe. Overall, IIoT Solutions are slated to improve competitiveness of the manufacturing sector. Figure 6 depicts a 3V Model comprising of Value Model, Value Chain and Value Proposition describing the Why, How and What?

- Value Chain – How does Industrial IoT Solutions transform Business?
- Value Proposition – What is the Impact of Industrial IoT Solutions?
- Value Model – Why should Manufacturing Organizations adopt Industrial IoT Solutions?

A survey conducted by Alphawise on behalf of Morgan Stanley to gauge people' expectations from Industrial IoT Solutions reveals the following benefits.

## INDUSTRIAL AUTOMATION USING IIOT

IIoT is basically a convergence of industrial systems (factories/plants, assemblies, machines) with advanced near real time computing and analytics, powered by low cost and low power sensing devices leveraging global (internet) connectivity. This

*Figure 6. 3V model (value chain, value model, value proposition)*

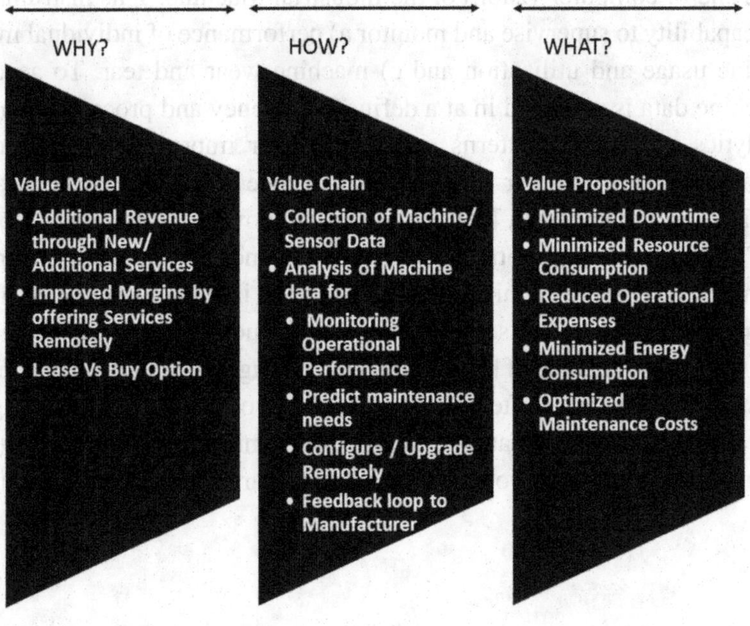

WHY?

**Value Model**
- Additional Revenue through New/ Additional Services
- Improved Margins by offering Services Remotely
- Lease Vs Buy Option

HOW?

**Value Chain**
- Collection of Machine/ Sensor Data
- Analysis of Machine data for
  - Monitoring Operational Performance
  - Predict maintenance needs
  - Configure / Upgrade Remotely
  - Feedback loop to Manufacturer

WHAT?

**Value Proposition**
- Minimized Downtime
- Minimized Resource Consumption
- Reduced Operational Expenses
- Minimized Energy Consumption
- Optimized Maintenance Costs

*Table 2. Survey results: benefits of industrial IoT solutions*

| SI No | Benefits | %Respondents |
|---|---|---|
| 1 | Improve Operational Efficiency | 47% |
| 2 | Improve Productivity | 31% |
| 3 | Create New Business Opportunities | 29% |
| 4 | Reduce Downtime | 28% |
| 5 | Maximize Asset Utilization | 27% |
| 6 | Sell Product as a Service | 18% |
| 7 | Reducing Asset Life Cycle Costs | 18% |
| 8 | Enhancing Worker Safety | 14% |
| 9 | Enhance Product Innovation | 13% |
| 10 | Better understanding of Customer Demand | 9% |

Source – Morgan Stanley Automation World Industrial Automation Survey, Alphawise

essentially means that one can supervise, control and alter the performance of an ecosystem of factories (multi-site plants/factories/ assembly lines) from either an individual factory or remotely anytime, anywhere vide. any-path /network.

## 1.   Smart Intelligent Machines

At the base of Industrial Internet of Things is an intelligent instrumentation (sensors, actuators and/or computer vision) of the Industrial Machine. This instrumentation imparts capability to supervise and monitor a) performance of individual machines b) machine usage and utilization and c) machine wear and tear. To accomplish this, machine data is streamed in at a defined frequency and processed using high end analytics to identify patterns and insights for improved decision making. This invariably leads to better machine performance at reduced costs with higher availability cum reliability. In IIoT terms, this is known as "Intelligent Optimized machine" – a machine that operates at peak performance and enables both operating and maintenance costs to be minimized. Further, the intelligence creates capability for the machine to become "self-acting" or "self-moving" or "self-configuring" machine that is also capable of being regulated, configured, and managed remotely.

Industrial IoT has thus shifted the traditional role of man(-mind) working on the machine to autonomous operations, by instilling 'mind' in the machine (locally, remotely or a combination of both). Smart Machines are capable of taking intelligent decisions based on defined rules and past historical machine performance – rules and high-end machine learning models are specified and work on the data collected and aggregated from sensors to predict and prescribe decisions. The decisions are relayed on to actutators which then implement the decisions – all this happens in an autonmous manner in a near real-time mode.

## 2.    Smart Intelligent e-Factories/ Plants

In a typical manufacturing system, individual connected Smart Machines combine together to create intelligent assembly lines which then combine together to create a set of intelligent Plants which when combined together create Connected Intelligent Factories also known as e-Factories. The connections are enabled by Machine-to-Machine (M2M) communication which creates a mesh of interconnected intelligent machines which balance, regulate and manage the operations. The Key Objectives of the e-Factories (refer to Figure 7 is to drive Industrial Productivity and conserve precious resources like energy.

Further, IIoT Solutions render Connected Intelligent factories/ e-Factories to collect, aggregate and analyze real-time data at both the edge level and remotely to rapidly detect quality faults and deviations (outliers, anomalies) which means corrective/ preventive actions can be taken immediately to reduce the downtime. Lastly, automated analysis of manufacturing process data is also being used to identify deviations in product quality. Thus, instead of a binary view (accept/ reject), the technique assists organizations in getting comprehensive insights into quality of production process and thereby help improve quality levels substantially.

*Figure 7. e-Factory objectives*

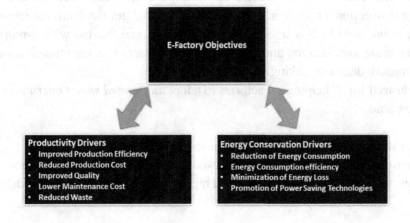

3.    Smart Configuration and Maintenance

IoT enabled Smart Machines have the capability to transmit operational and performance data information to operators, field engineers and original equipment manufacturers. This coupled with High-end data Analytics is used to monitor the condition and performance of the Smart Machines. Also termed as "Condition Based Monitoring", this essentially means that the condition and performance of machine is continuously monitored and proactive actions taken as soon as machine performance degrades. IIoT is thus helping in maximizing the utilization and performance of the Machines. Additionally, the aggregated time-series data coupled with images (e.g. wear and tear images) is correlated and analyzed to detect patterns and new insights in terms of potential future failures and the spare parts required for maintenance. Accordingly, the maintenance process is being transformed from being reactive (i.e. Scheduled or Breakdown maintenance) to proactive and predictive – thereby, reducing both the maintenance costs and maximizing machine uptime and availability.

This is a marked improvement from traditional manufacturing, where-in discrete machine condition measurements are taken periodically by workers and analyzed for deviations. The maintenance process in traditional manufacturing is largely preventive (scheduled) and/or breakdown (reactive). With Smart Machines, the process becomes essentially condition based and proactive. Further, as against manual inspection for wear and tear of rotating parts e.g. drills, computer vision-based inspections and analysis can further automate decision making and improve the product quality.

The automated process also acts like a feedback loop and the field engineers/ experts can remotely diagnose/ supervise the machine usage and reconfigure as necessary thereby improving the machine performance and availability. Additionally, remote monitoring and configuration reduces the need to have experts/ specialized technicians located in the factory premises. Experts/specialized technicians can now be centrally co-located at one place i.e. at an Operational Command Centre and can monitor and reconfigure (using actuators) multiple connected e-Factories remotely. From a conventional one-man: one-machine the paradigm now shifts to one-man: multiple machines and/ or one-man: multiple plants and that too with appropriate context aware visualization and descriptive cum prescriptive recommendations to help improve decision making.

Industrial IoT is helping e-Factories to adopt *automated smart operations* that are now able to:

- ensure prescribed working environment for machinery
- ensure prescribed diagnostics are used to vet the machine performance
- ensure prescribed maintenance and repair processes are executed on demand.
    4.    Smart Workplaces

In traditional manufacturing, factory worker' Health, Safety and Environment (HSE) issues are always major big ticket items on management agenda. Often the major reason for worker dissatisfaction and absenteeism is exposure to unhealthy / hazardous working environment. This is especially true in heavy industries like Steel, Oil and Gas, Mining etc. Today, IoT devices (mostly wearables) are being used to automate and track worker locations and monitor vital parameters (like blood pressure, heart rate, etc.). This automated effective monitoring helps to reduce the number of injuries and illness, near fatalities, absences both short and long term during daily operations and improves work force morale and motivation.

Further, with environment sustainability gaining popularity, manufacturing industries are increasingly being subjected to strict regulatory compliances. IoT devices are helping manufacturing companies in collecting and analyzing raw environmental data to take effective steps as both remedial and preventive mechanisms. Finally, the data trail for all HSE practices is being used to file Sustainability reports and is routinely used during auditing function. Thus, automation of HSE practices is being enabled by IoT devices which ensures proper collection, dissemination and auditing of health, safety, and environment (HSE) issues.

5.    Smart Supply Chain

Going up-stream, IoT Industrial applications permit monitoring of events across the entire "supply chain". Using these systems, both the raw material and finished goods inventory can be tracked and traced globally across all the warehouses at a line-item level and the management teams notified of any deviations from the agreed plans. This real time visibility of inventory (raw, finished goods) across the supply chain can help optimize the supply chain. Progressive manufacturing companies are connecting their e-Factories to suppliers, and other third parties concerned with supply chain to trace interdependencies, regulate material flow to just-in-time and improve manufacturing cycle times. On the other side, manufacturers are giving visibility to finished goods inventory and work-in-progress purchase orders to customers to help manage demand better through transparency (Martin, Thilo, & Jurgen, 2017). Essentially this automation is helping manufacturers to move from a "Make to Stock" to "Make to Order". Figure 8 depicts the Supply Chain automation using Industrial IoT.

For outbound logistics, fragile high value products are now increasingly being tagged with smart sensors to track and trace their real-time location, ambient environment and exposure to shock absorptions/ vibrations and its impact on product during transit. This is especially true for pharmaceutical and food and beverages industry where the ambient environmental conditions are vital to protect and manage product' integrity.

*Figure 8. Supply chain automation using industrial IoT*

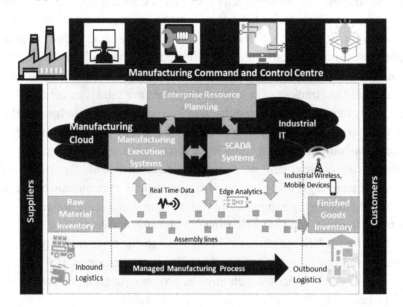

Looking inwards, IoT sensors are being deployed to monitor the manufacturing process starting from raw material refining process down to the packaging of final products. This complete end-to-end monitoring and automation of the manufacturing process in (near) real-time provides visibility to factory management to adjust operations as needed thereby improving utilization and reducing operational costs. The close monitoring also provides visibility to work in progress inventory thereby giving vital inputs for automating the inventory replenishment process. This is a step-improvement from manual inventory management processes where Purchase orders are filled in manually by workers depending upon the manufacturing progress.

## 6. Multi-Site Industrial Automation

Today, global Multi-national companies (MNCs) have multiple production facilities spread across the globe for the same product. Figure 9 depicts the Multi-site Industrial Automation where in a Global Manufacturing Command and Control Centre acts as the "nerve" centre controlling the Order Management, Production Planning and Operations, Plant Maintenance, procurement, and overall Management. This enables all the core processes to mature and truly become "Global Processes"

*Figure 9. Multisite industrial automation*

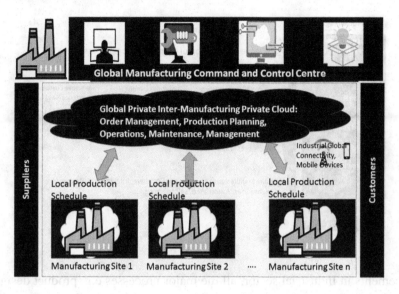

with Standard Operating Procedures. The standardization provides much needed flexibility and dynamism to react quickly to changing market conditions. This is a classic scenario of centralized management with decentralized operations (command and control structure). For example, all customer orders are processed centrally and routed to local manufacturing sites for production. Similarly, all procurements are processed centrally and purchase orders are sent to suppliers to replenish raw material at local manufacturing sites. All the key metrics are managed centrally with distributed operations. This allows organizations to exploit the local advantages with global expertise to develop products cheaper, faster and better.

A case in point is a Project codenamed GRACE – inteGration of pRocess and quAlity Control using multi-agEnt technology (Paulo, Armanda, & Stamatis, 2016) being piloted in Europe. The key aim of the pilot is to enable factory owners to integrate Global and Local Schedules to improve efficiency of production, product customization and quality control.

## 7.    Smart Manufacturing Process

As discussed above, Industrial IoT has impacted all stages of manufacturing right from optimizing machine performance, to improving inventories and making supply chain more effective, to improving worker' health and safety. With technological

*Figure 10. Manufacturing process transformation using digital technologies*

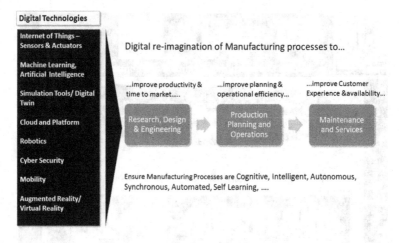

advancements in Industrial sector, all the major processes e.g. product design and engineering, production planning and operations, maintenance and services production are now simultaneous activities that are performed by high performance teams in an integrated fashion. This is a marked improvement from the serial process flow in the conventional manufacturing organizations. Figure 9 depicts the transformation of Manufacturing Processes using Digital Technologies to make manufacturing processes Cognitive, Autonomous, Synchronous, Automated and Self learning (Siemens, 2017). Finally, progressive manufactures who invest in building "product digital twins" as part of their Design and Engineering process [digital twin are virtual intelligent models that simulate real-world characteristics and performance of actual physical manufacturing sites] can use IoT based data to further refine the "Design and Engineering" process to improve product features, specifications and quality. This gives the manufacturers an ability to virtually simulate the real-world scenarios at a fraction of cost.

8.    Smart Collaboration with Partners

IIoT enables sharing of real time machine performance data with Original Equipment Manufacturers (OEMs). This sharing acts as a feedback loop and enables deeper understanding of actual machine usage and performance in real time operating environment to OEM Designers, assisting them in improving the quality and useful life of machine. Further, this is assisting OEMs to move from

Product-Centric approach to Customer-Centric (Erhad, Markus, Rainer, Daniele, & Jorg, 2016). IIoT has started to also re-shape "New Collaborative Business Models" where the OEMs are moving from machine sellers to sellers of "machine services" vide service contracts.

This model is a win-win for both OEMs and factory owners as OEMs get machine usage and utilization data along with long term customer commitment and for factory owners the procurement costs transforms from Capital Expenditure (Capex) to Operational Expense (Opex). Figure 11 depicts the IoT Services Roadmap Vs Revenue Expansion through maturing OEM- Factory Owner Engagement (Raman & Rob, 2017). In the new model, the factory owner now leases the machines and the machine life cycle management is performed by the OEMs.

In a nutshell, Automation using Industrial IoT is helping manufacturers to:

- Standardize performance and service levels (across all stages), thereby improving reliability
- Eliminate uncertainty in response times (rule-based decision making)
- Reduce human errors
- Improve health and safety of workers.

As discussed above, the Optimization Opportunities are immense and spread across the entire Manufacturing landscape. Figure 12 shows the macro level Optimization Opportunities using Industrial IoT across Business, Operations and Machine performance.

*Figure 11. IoT services roadmap vs revenue expansion*

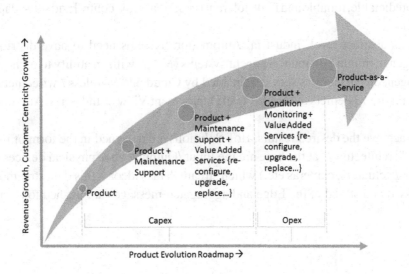

*Figure 12. Optimization opportunities using industrial IoT*

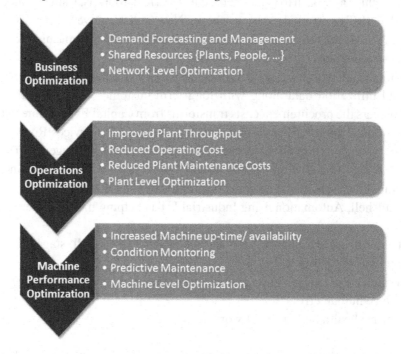

## IIOT SOLUTIONS: DEPLOYMENT ARCHITECTURE

Industrial IoT Solutions are usually large and complex systems. From an architecture perspective, Industrial IoT Solutions require secured scalable collaboration between various kinds of devices and systems, intelligent services at the edge and cloud to offer predictable, reliable and fault-tolerant (near) real-time control and self-adapting systems.

At an abstract level, Industrial Automation Systems need to have distributed intelligence running on multiple agent systems (MAS) with an ability to orchestrate heterogeneous devices as services enabled by Cloud and Wireless / wired network mesh. Figure 13 depicts the High-Level Deployment View of Industrial Automation Systems.

To achieve the desired results, IIoT Solutions are designed in the form of multi-layered architecture – at the base are the Input /Output Cyber-physical devices (IoT Sensors/ actuators, cameras etc.) which stream data/ videos – this data streaming is processed and stored at the Edge and aggregated messages are pushed to complex

*Figure 13. High-level deployment view of IoT solutions*

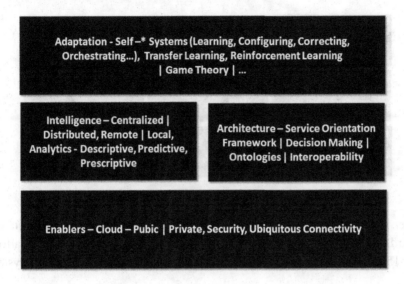

event processing layer which is responsible for processing multiple events. Processed messages are stored in data lakes/data bases and integrated with Enterprise Resource Planning data [e.g. Orders, Production Plan, etc.]. This data is analyzed at the Analytics layer using Business Intelligence and Machine learning models and presented in the form of application level dashboards, heat-maps and other visualization maps to relevant Users for controlling and managing operations. Figure 14 shows the Functional Capability Map across the different layers of the Industrial IoT Solutions.

*Figure 14. Capabilities map*

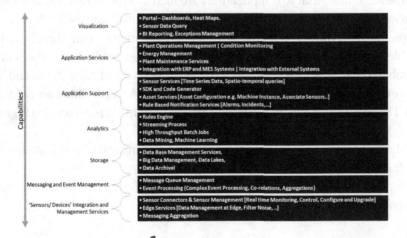

*Figure 15. IoT standards and ecosystems*

| Industry & Government | Competing Ecosystems | Certifications |
|---|---|---|
| • Industrie 4.0,<br>• Smart Grid,<br>• Smart Infrastructure, ... | • Google, Apple, Amazon,<br>  Siemens, Bosch,<br>  Samsung, ... | iCSA Labs, UL, ... |
| **Industry Consortiums** | **Protocols and Technologies** | **Monitoring** |
| • Industrial IOT<br>  Consortium, Open Fog,<br>  M2M, ... | • MQTT, CoAP, DDS, AMQP,<br>  XMPP, ... | SNMP, REST, BCI, .... |

**Networking**
• Near Field – Bluetooth, RFID, ANT, 802.11
• Campus/ Region – Zigbee, Bluetooth, 802.11, UWB, LoRa, SigFox
• National/ Global – LTE, 5G, Satellite

Over the last few years, there have been significant efforts to usher in set of Standards and Guidelines to streamline development and adoption of IoT Platforms. Figure 15 shows the current state of the evolving IOT Standards and Ecosystems.

## IIOT SOLUTIONS: CHALLENGES AND RECOMMENDATIONS

Industrial IoT Solutions are large complex systems with significant investments and while they are very beneficial in terms of automation, quicker time-to-market, workforce safety and cost reduction they do present a number of issues and challenges that need to be addressed to get sustained return on investments (Hongyu and Kristian, 2015).

1.    Security

Large scale Industrial IoT applications are increasingly vulnerable to industrial espionage i.e. disruption from malicious attacks or information theft. More devices, systems and Machine-2-Machine communication means that there are today more 'decentralized' entry and exit points – the bigger the footprint the more the vulnerability. Industrial IoT Systems have multiple layers of abstraction and hence there are multiple attack surfaces as enumerated below (Ahmad, Christian, & Michael, 2015). The Security Vulnerability of Industrial IoT Solutions is depicted in Figure 16.

a)    Machines /cyber-physical systems electronics can have physical attacks, side-channel attacks.
b)    Software at edge level and cloud level can be infected by viruses, trojans and runtime attacks.

*Figure 16. Security vulnerability of industrial IoT solutions*

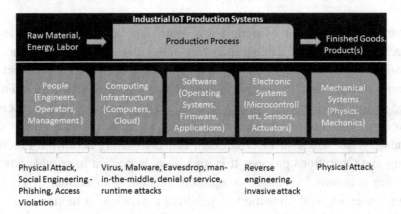

c) Communication protocols can be attacked by man-in-the-middle and/or denial of service attacks, and finally

d) Humans controlling the above can be subject to social attacks (phishing, social engineering).

To safeguard from above, the IIoT Solutions need to have end-2-end security including securing the devices, data at motion and at rest and client/ end-user security. This is enabled by provisioning advanced security mechanisms such as secured and resilient layered architecture, multi –level encryption and user authentication / authorization and a robust threat detection-cum-management system. This can be broadly classified as below:

a) Endpoint Security – This includes device hardening, device authentication, encryption of data during transmission (network transport encryption), secure key storage etc.

b) Communication Security – Server authentication, Channel encryption, Secure Messaging gateway and payload encryption

c) Service Security – Data at rest encryption, API Gateways, Fine grained access controls, policy based access controls and whitelisting/ blacklisting of client endpoints

It is important that the IIoT Solution are regularly subjected to Security Vulnerability Assessment and that Security audits are carried out at periodic intervals. As a best practice it is recommended to deploy a Security Operations Centre as part of the Overall Operations Command Centre.

2.    Data Integrity, Flow and Management

**Incorrect Data and/or Missing Data**: With data being the fuel for decision making, unavailability of data or incorrect/ noisy data or out of sequence data can have severe consequences on Industrial IoT Solutions as they can render the entire analytics platform ineffective leading to poor end user satisfaction. Further, in some sectors this can lead to compromised national security, this is essentially true for large scale utility companies e.g. Power Systems/ national grids, Oil and Gas Industry etc. where-in incorrect data can completely derail the Production Planning and Operations process. It is, therefore, important to provide advanced networking solutions to manage congestion, reliable transmission of encrypted data, and an end-to-end protection of sensitive data while at motion and at rest. Also, as Industrial IoT devices are interconnected, the users must have the trust and confidence that the data is fresh (i.e. relevant) and that the data mapping and monitoring rules and information being exchanged are both accurate, safe and can be relied on. It is imperative that the right data is available at the right instance for right decision making. Further, the actions of the IoT devices (sensors, actuators) should be non-repudiating (Shahid, Ahmed, Zhibo, Ammar, Kim, & Jonathan, 2017) i.e. the devices cannot deny the actions after being performed. This is essential to ensure integrity of Industrial Operations. In order to manage the same, BlockChain technology is being experimented to create a distributed ledger of operations. The fundamental benefit that BlockChain provides is immutability and a distributed view of operations.

As discussed above, data is the soul of Industrial IoT Solutions and it is an imperative that the data is collected, transmitted, computed, stored and archived as necessary. Industrial IoT data is essentially "Big Data" as it is voluminous, comes at a high velocity and has multiple varieties. Figure 17 depicts the Industrial Automation Data Flow and Databases that are created and maintained to continuously improve and optimize operations. It is important to note the integration needs and processing of Enterperise Resource Planning/ Manufacturing Execution Systems data with real time data collected from Sensors to improve the operations and performance. Further, Advanced Machine learning algorithms/models are being used to create intelligent models to predict and prescribe optimized performance. This is in stark contrast to the conventional knowledge management in manufacturing sector, where in the plant process data is maintained within the confines of the plant itself and the analytics, if any, is carried out for local learning and feedback.

*Figure 17. Industrial Automation Data Flow and Databases*

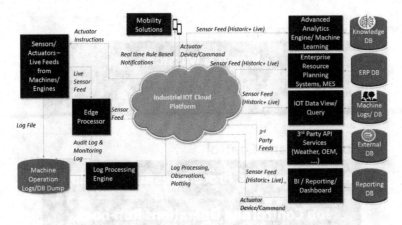

## 3. Interoperability

Industrial IoT Solutions need to operate with a large number of heterogeneous devices (large variety from different suppliers), legacy operations technology (SCADA – Supervisory Control and Data Acquisition Systems) and manufacturing execution systems. Today, the manufacturing industry is fraught with proprietary interfaces and solutions. This essentially means that there are usually multiple platforms, numerous protocols and large number of Application programming interfaces. While standards have started to emerge, the adoption is both time-consuming and costly and there is a need to bridge the existing gap.

Figure 18 depicts the various layers of interoperability (Daryll et al., 2015) requirements for IIoT Solutions. At the source, given the heterogeneity of devices there is a need to have physical interoperability – i.e. the various types of devices should be able to communicate with each other. Today, the issue is that the same type of device [by different providers] give data in different formats, quality and frequency thereby making data pre-processing a key activity before feeding the data into a Data lake/ Database. The next layer is the interoperability at the Network layer to address the variation in the current protocols say across multiple sites in a Multi-site Solution or multiple protocols within the same site i.e. different protocols for different assembly lines. At the Data and Information layer, both the Data Model and Information Model should be flexible enough to transform and create uniform information layer for extraction of knowledge – a simple example being

*Figure 18. Interoperability layers*

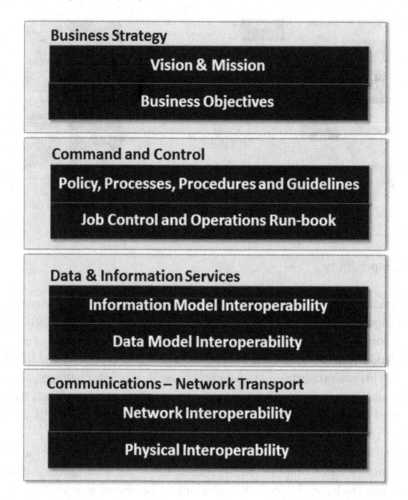

harmonization of unit of measurements, or say frequency of data collection from the same type of machine across multi-sites. It is important to create a uniform data profile which acts as a source for feature engineering to build Machine learning Models. The Dataset preparation and validation is an important step in Machine learning Analytics. Interoperability is today mandatory feature for any large scale Industrial IoT solutions architecture.

4.    Latency

Industrial IoT applications have literally hundreds of thousands of sensors, actuators, PLCs connected – it is therefore important that the data transfer and analysis is near or real-time. Due to the use of wireless internet connections, there is an issue of latency that needs to be managed in latency sensitive environments – this usually results in deploying data and service intelligence at different levels from network edge (localization of computation), to local cloud, and finally to centralized cloud. Figure 19 depicts the Edge Computing Architecture comprising of Edge Analytics, local memory and device management agents running on a Lite Operating System with a Lite Database. The parameters that impact the latency are bandwidth availability, data location, overheads, errors etc.

Today, software defined IIoT architectures inclusive of software defined networks are being deployed to manage physical devices and interfaces for information exchange (Jiafu et al., 2016).

One of the key concerns in Industrial IoT solutions is the latency and data accuracy/ precision as small fluctuations in reported vs actual data can severely jeopardize the operations. It is also equally important that the data arrives in the right sequence/ events in order to take the right responses. Industrial operations usually need higher levels of precision, accuracy and need to be synchronized to usually milliseconds. Normally, stringent Quality assurance processes are deployed

*Figure 19. Edge computing architecture*

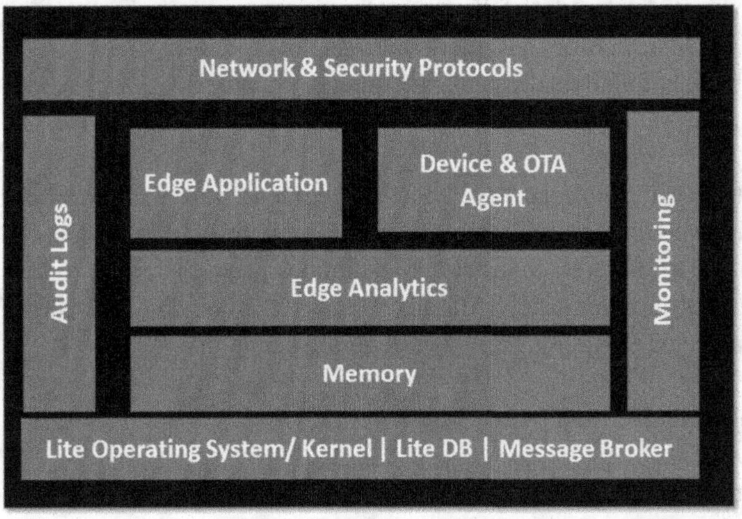

to detect minor deviations and take immediate corrective actions. The parametric analysis of real-time systems generally use Parametric Timed Automata which is an extension of timed automata (Thi, Luigi, Roberto, Yusi, & Alessandro, 2010), with inclusion of parameters and state variables.

5.    Scalability, Reliability and Availability

Industrial IoT applications need to be scalable at multiple levels – starting from scalability from one individual machine with one or more devices to multiple machines to multiple connected factories. Scalability, therefore, has multiple dimensions:

a)    At a device level, the naming convention and address space must be scalable to add more devices;
b)    Data communication and networking level should be scalable to connect new devices and new networks without impacting the existing network; and
c)    Service management and provisioning – lastly the applications have to be scalable to manage a large number of geographically spread disparate connected factories running heterogeneous devices on multiple cloud infrastructures.

Further, Large scale Industrial IoT applications need IoT devices and connections that are both reliable and available when demanded – the main parameters that are used to track reliability and availability are:

a)    MTBF – Mean time between failures- likelihood of devices, set of devices or a connected network to fail
b)    MTTR – Mean time to Repair – average time to repair a fault
c)    Probability to fail on demand – likelihood of a process, device working as per the need and thereby the likelihood of failure of device or process

Industrial IoT Solutions have long replacement time windows i.e. they are supposed to work in harsh environments usually remote for long periods of time with precision and no failures. Industrial IoT solutions need, therefore, to be hardened and have built-in redundancy (Masato, Yosuke, Tomini, & Shuji, n.d.).They are usually subjected to rigorous tests before implementation.

6.    Fault Tolerance and Safety

Industrial IoT solutions consist of thousands of heterogeneous industrial devices, therefore, it is vital to design fault tolerant and reliable systems to adapt to various scenarios like service failures, malicious attacks, failures in cloud or network infrastructure etc. The key aim is to develop resilient systems that are able to recover from failures or fail gracefully by contextually switching the load to available capacity. IIoT solutions should be capable of balancing load, and have adequate redundancy to ensure fail safe operations.

Industrial IoT solutions also need to adept to stringent functional safety standards and regulations as required by the Industry – this is usually achieved by correct execution of commands and functions. Multi-agent Industrial IoT solutions also need to conform to country specific regulatory compliance needs. Safety is a broad topic, but for Industrial Automation systems both intrinsic safety (e.g. electric/ electronic failures leading to fires) and functional safety (failure in the system should not lead to dangerous/ hazardous consequences) are important and relevant. Most of the industrial networks today support functional safety protocols i.e. automation system is kept at a safe state at all times to prevent failures and malfunctions (Tomas, Mikael, & Johan, 2017).

## 7. Maintainability, Serviceability and Programmability

Industrial IoT Solutions are supposed to perform reliably for long periods of time with minimal or low downtime. Hence, it is important that the IoT solutions have certain amount of redundancy built-in and that they are serviceable with all types of maintenances – breakdown, preventive, adaptive and predictive. Given the fact that the Industrial IoT Solutions have diverse components spread across multiple layers in terms of hardware [devices and networking], firmware, system software, databases and application software and are rendered by multiple service providers it is important that the maintenance needs are proactively managed as any fault in any component or layer has the potential to bring down the entire Solution. Industrial IoT Solutions are more often than not mission critical solutions supporting core business processes, hence, it is imperative that the solutions are highly available and maintainable solutions with near zero down time. Figure 20 shows the Hire-to-Retire process for deployment of Industrial IoT Solutions.

As the IIoT Solutions are very large and diverse, with multiple service providers, there is usually a Guardian Vendor, who co-ordinates and orchestrates the maintenance activities across multiple service providers. Guardian Vendor ensures that the Maintenance services are as per stipulated Service Level Agreements backed by relevant Operational level agreements with Service providers. Further, the Maintenance services for Multi-site Industrial IoT Solutions are provisioned both at the local [say industrial site] and global level and follow the sun approach in terms of maintenance activities.

*Figure 20. Hire-to-Retire process*

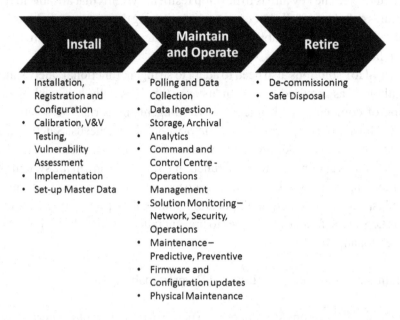

Industrial IoT solutions also need to have the flexibility to be re-programmed /re-configured locally or remotely to support new adaptation/processes. The re-programming may be required at multiple levels - at the edge level, devices should be capable of re-configuring/ re-booting either locally or remotely. At the Edge software level, the reprogramming could be installation of a new version of analytics software or a new version of the Lite Operating System or the Database. Similarly, at the platform level, re-programming could mean adding new intelligence services, and an ability to modify or change the system software. Essentially the IIoT Solution should be capable of re-programming at multiple layers and across multiple components. The programming is usually done remotely, although the same may be rendered from on-premises.

8.    Workforce Talent

Industrial IoT Solutions are complex and need a vast array of skills and competencies ranging from Instrumentation/electronic engineers, Network engineers, Data scientists, Computing infrastructure (Cloud, Operating System, Edge), Visualization experts (User Interface Designers), ERP specialists, Industrial Domain Subject Matter

Experts etc. These skills are required both during the development and deployment of IIoT Solutions as well as during the normal operations. Assembling, motivating and keeping such talented workforce engaged at all times is an issue. In order to manage the risk of attrition and its impact on operations, most of the organizations maintain a certain amount of redundancy which results in escalated costs. Further, factory operators, supervisors and management team i.e. the users and beneficiaries of the Industrial IoT solutions need extensive training to instill knowhow in terms of usage and performance. While the major advantage of automation is increased machine-man ratio, the disadvantage is over-dependence on the few men that are required to run the operations. Lastly, deployment of Industrial IoT Solution is invariably a large Organization Change Management Program and needs extensive leadership guidance to make it effective.

9.    Geo-Political Landscape

Multi-site Industrial IoT Solutions are usually spread across the globe i.e. e-factories are spread across multiple countries and are based on the principle of aggregated centralized decision making to generate economies of scale. Emergence of restrictive trade and data protection practices in many countries is fast becoming a major issue. In recent years, there is an emerging trend to impose trade tariffs to protect local industry from global competition as also the requirement to process and store critical data locally. Secondly, an inherent assumption of Multi-site IIoT solution is a global specialized workforce, again with stringent restrictive visa norms global mobility is at risk. Geo-Political stability is a core fundamental requirement for successful running of Multi-Site Industrial IoT based e-Factories.

# CASE STUDY: OIL AND GAS INDUSTRY

"Digital Rig" in Oil and Gas Industry is a great example of application of Industrial Internet to achieve productivity gains and optimization. The key aim of Oil and Gas Industry, today, is to:

- Improve operational efficiency and productivity
- Lower life cycle costs
- Improve safety and environmental/ regulatory compliance
- Refurbish aging facilities
- Support remote locations

*Figure 21. GE digital rig- globetrotter*
Photo source: Globetrotter Vessel I, GE, Noble Corp

Oil and Gas Industry can leverage IoT devices for:

- Deployment of downhole sensors to track oil-well events, rotating machinery, pressure sensors in pipes for optimization of oil flow vide Wireless Communication Systems
- Predictive and prescriptive analytics to better understand / anticipate reservoir characteristics
- Monitoring e.g. 4-d seismic, fluid migration and reservoir changes
- Real-time data monitoring for safety and optimization

Example – Noble Globetrotter I Vessel – World's First Digital Rig ("GE Digital Rig")

Noble Corporation, an offshore drilling contractor and GE have partnered to create world's first digital drilling vessel – this is a truly innovative step to enhance offshore marine operations using digital technologies. In order to do so, GE Digital has developed Digital Rig$^{SM}$ solution, powered by its GE Predix Platform. With this, Noble is able to:

- Enhance data-driven operations support
- Gaining visibility into drilling inefficiencies
- Connect all the control systems viz. drilling control network, power management system, dynamic positioning system, etc.

Device Data is collected vide individual sensors/ control systems, filtered, harmonized / centralized (edge) on the Rig vessel before transmitting in (near) real-time to GE's Industrial Performance & Reliability Center for analytics. Noble is also leveraging GE's digital twin technology to create a digital replica of the physical asset. This shall enable Noble to perform advanced analytics to detect anomalies, and offer early warning/ alert/ alarm signals to equipment operators, who can then mitigate the issues proactively. This shall enable personnel both on-board (vessel) and onshore to gain a 360-degree view of the vessel's health status and real-time performance status of each onboard equipment.

The technology has shown promising results, it has been able to capture likely major anomalies, and produce alerts to inform potential failures up to two months in advance. Further, GE and Noble have set a target to achieve 20% reduction in operational expenditures on specific targeted equipment. Lastly, thru leveraging digital twin technology, Noble's personnel on the vessel can now focus on maintenance activities that are needed and are most effective—this is helping Noble to reduce unplanned downtime, thereby, improving revenue and generating significant maintenance cost savings.

The 'digital spine data' backbone implemented on the vessel is a great source of building valuable data repository which shall open the door to transforming additional rigs for Noble and eventually pave the way towards autonomous drilling and unprecedented asset performance improvements.

## FUTURE RESEARCH DIRECTIONS

Industrial Internet of Things is a core emerging technology aimed at revolutionizing the manufacturing industry by building intelligence at each layer to improve automation, productivity, quality and safety. Though a proven concept, the adoption has been slow but is likely to gather pace in future. IIoT applications are large complex systems and comprise of heterogeneous objects – there is a need for this objects to be able to communicate seamlessly – though Standards have started to emerge across all the layers, there is considerable scope of research in this area. Besides this, given the nature and the stage at which the maturity of Industrial IoT applications is today, there are a number of areas where significant amount of research is happening or shall happen in future, as given below

- **Networking**: Wireless transmission, networking protocol, congestion management, improve transmission speed and data encryption and compression/ de-compression, M2M communication.
- **Device Management and Control Systems**: Device Monitoring, Device upgrade, Event processing at Edge, legacy devices.
- **Security Key Focus Areas**: Physical, Cybersecurity [data at capture, motion and rest], certificate management, Database encryption.
- **Data Quality and Analytics**: Data Streaming and Interoperability, Transformation, Management, Feature Engineering, Analytics – Neural, Deep learning and Machine learning Models, Performance Management, Data Archival.
- **Domain**: Industry Domain, Digital Twins to simulate product performance.

## CONCLUSION

The Industrial Internet of Things is a network of physical objects/machines, computing systems and digital technologies, platforms, ubiquitous networks and applications that have embedded technology to share and communicate intelligence with each other, and also with the external environment inclusive of people. The widespread proliferation of IIoT is being enabled by availability and affordability of sensors, computing resources (e.g. processors) and technologies like Cloud, Big Data and Machine learning backed by fast and reliable networks. The computing environment facilitates capture of and access to real-time information for decision making. IIoT, today, is helping industries to improve productivity and operational efficiency, reduce operating costs and enhance worker safety. IIoT also has the capability to create new revenue streams by transforming organizations from being product-centric to customer-centric. IIoT adoption is fast emerging as a key driver in transforming and automating manufacturing processes.

Though very promising, the adoption rate is still slow. Only 7% of the Industries surveyed have adopted large scale IIoT. Despite automation, the net impact of IIoT is going to be net positive in terms of job creation – intelligent machines are likely to automate mundane tasks, this will un-lock the potential of workers to do more creative and collaborative work. IIoT Solutions are large and complex and span across major Business processes, hence it is essential that these systems are designed for high availability, low latency, high redundancy and reliability, resilience and safety. As with any large system, IIoT Solutions need to be protected both physically and for cyber threats and should be continuously monitored for vulnerabilities. For large scale global businesses, journey towards adoption of IIoT really translates into 3 major aspects a) Create new Industry Models b) Harness and capitalize the value of data through descriptive, predictive and prescriptive analytics c) Build talent and workforce ready for the future!

## REFERENCES

Amer, W., Ansari, U., & Ghafoor, A. (2009). Industrial automation using embedded systems and machine-to-machine, man-to-machine (m2m) connectivity for improved overall equipment effectiveness (OEE). In *Proceedings of 2009 IEEE International Conference on Systems, Man and Cybernetics,* pp. 4450-4454.

Annunizata, M. & Bell, G. (2016). *Digital Future of the Electricity and Power Industry*, GE Power Digital Solutions, GE

Biswas, D., Ramamurthy, R., Edward, S. P., & Dixit, A. (2015). The Internet of Things: Impact and applications in the high-tech industry, Cognizant 20-20 Insights, Cognizant Technology Services, New Jersey. Available at https://www. cognizant. com/whitepapers/the-internet-of-things-impact-and-applicationsin-the-high-tech-industry-codex1223. pdf.

Chitkara, R., & Mesirow, R. (2017). *The industrial internet of things.* London, UK: PricewaterhouseCoopers.

Colla, M., Leidi, T., & Semo, M. (2009), Design and implementation of industrial automation control systems: A survey. In *Proceedings of 7th IEEE International Conference on Industrial Informatics* 10.1109/INDIN.2009.5195866

Consel, C. & Kabac, M. (2017). Internet of things: From small- to large-scale orchestration. In *IEEE 37th International Conference on Distributed Computing Systems.*

Conway, J. (2015). The industrial internet of things: An evolution to a smart manufacturing enterprise. *Schneider Electric.*

Evans, P. C. & Annunziata, M. (2012, November). *Industrial internet: Pushing the boundaries of minds and machines*, GE Digital, GE.

Fiege, E., Hammer, M., Ulrich, R., Iacovelli, D., & Bromberger, J. (2016). *Industry 4.0 at McKinsey's model factories.* McKinsey Corporation GE, Digital Rig, Retrieved from https://www.ge.com/digital/blog/worlds-first-digital-rig-digitizing-operational-excellence

Fogal, D., Rauscheker, U., Lanctot, P., Bildstein, A., Burhop, M., Caneodo, A., … Xiaonan, S. (2015). Factory of the future. *IEC.*

Harris, K. *(2008, September). An application of IEEE 1588 to industrial automation. In 2008 IEEE International Symposium on Precision Clock Synchronization for Measurement, Control and Communication (pp. 71-76). IEEE.*

Kaiza, J. (2016). A new age of industrial production The Internet of things, services and people. Zurich, Switzerland: ABB Group.

Khan, M., Wu, X., & Dou, W. (2017), Big Data Challenges and Opportunities in the Hype of Industry 4.0, Big Data Networking Track, *IEEE ICC 2017 SAC Symposium*

Leitao, P., Colombo, A. W., & Karnouskos, S. (2016). Industrial automation based on cyber-physical systems technologies: Prototype implementations and challenges. *Computers in Industry, 81,* 11–25. doi:10.1016/j.compind.2015.08.004

Lennvall, T., Gidlund, M., & Akerberg, J. (2017). *Challenges when bringing IoT into Industrial Automation.* IEEE Africon. doi:10.1109/AFRCON.2017.8095602

Low, K.-S. & Keck, M.-T. (2003). Advanced precision linear stage for industrial automation applications. *IEEE Transactions on Instrumentation and Measurement, 52*(3).

McLaughlin, P., & McAdam, R. (2016). *The undiscovered country: The future of industrial automation. Honeywell Process Solutions.* Honeywell.

Mumtaz, S., Asohaily, A., Pang, Z., Rayes, A., Tsang, K. F., & Rodriguez, J. (2017). Massive internet of things for industrial applications. *IEEE Industrial Electronics Magazine,* 28–33.

O' Gorman, A. (2016). *Internet of Things in the Industrial Sector.* IBM Global Business Services, IBM Corporation.

Pei-Breivold, H., & Sandstrom, K. (2015). Internet of things for industrial automation-Challenges and technical solutions. In *Proceedings of IEEE International Conference on Data Science and Data Intensive Systems.* 10.1109/DSDIS.2015.11

Perera, C., Liu, C. H., Jayawardena, S., & Chen, M. (2014). A survey on internet of things from industrial market perspective. *IEEE Access: Practical Innovations, Open Solutions, 2,* 1660–1679. doi:10.1109/ACCESS.2015.2389854

Sadeghi, A.-R., Wachsmann, C., & Waidner, M. (2015). *Security and Privacy Challenges in Industrial Internet of Things, DAC.* San Francisco, CA: ACM.

Siemens. (2017). *MindSphere The Cloud-based, open IoT operating system for digital transformation.* Siemens PLM Software

Sniderman, B., Mahto, M., & Cotteleer, M. J. (2016). *Industry 4.0 and manufacturing ecosystems Exploring the world of connected enterprises.* Deloitte, USA: Deloitte LLP Consulting.

Thi, T. H. L., Palopoli, L., Passerone, R., Ramadian, Y., & Cimatti, A. (2010). *Parametric analysis of distributed firm real-time systems: A case study.* Piscataway, NJ: IEEE.

Wan, J., Tang, S., Shu, Z., Di Li, S. W., Imran, M., & Vasilakos, A. V. (2016). Software-defined industrial internet of things in the context of industry 4.0, IEEE Sensors Journal, 16(20).

Wikipedia. Internet of Things. Retrieved from https://en.wikipedia.org/wiki/Internet_of_things

Wollschlaeger, M., Sauter, T., & Jasperneite, J. (2017). *The future of industrial communication automation networks in the era of the internet of things and industry 4.0. IEEE Industrial Electronics Magazine.*

*Yamaji, M., Ishii, Y., Shimamura, T., & Yamamoto, S. (n.d.). Wireless sensor network for industrial automation.* Tokyo, Japan: Ubiquitous Field Computing Research Centre, Yokogawa Electric Corporation.

## ADDITIONAL READING

Automation, R. (2017). Unlocking the value of your Connected Enterprise, Digital Transformation in the Fourth Industrial Revolution, Retrieved from: http://www.rockwellautomation.com

Daugherty, P., & Berthon, B. (2015). *Winning with the Industrial Internet of Things.* Accenture.

Edson, B. (2014), Creating the Internet of Your things, Microsoft Corporation Charles Consel, Milan Kabac (2017), Internet of Things: From Small- to Large-Scale Orchestration, IEEE 37th International Conference on Distributed Computing Systems Daniela Lettner, Florian Angerer, Herbert Prahofer, Paul Grunbacher (2014), A Case Study on Software Ecosystem Characteristics in Industrial Automation Software, ICSSP, ACM

Frohm J, Linstrom V, Winroth M, Stahre J (2206), THE INDUSTRY'S VIEW ON AUTOMATION IN MANUFACTURING, IFAC

Gronau, N., Ullrich, A., & Teichmann, M. (2017). Development of the Industrial IoT Competencies in the Areas of Organization, Process, and Interaction based on Learning Factory Concept. *Procedia Manufacturing*, *9*, 256–261. doi:10.1016/j.promfg.2017.04.029

Jekielek, J. Managing Information on Industrial Automation Projects, *6th International Conference on Computer Information Systems and Industrial Management Applications*, IEEE

Miyachi, T., & Yamada, T. (2014), Current issues and challenges on cyber security for industrial automation and control systems, *SICE Annual Conference*, 2014 10.1109/SICE.2014.6935227

The Industrial Internet of Things (IIoT): the business guide to Industrial IoT. Retrieved from: http://www.i-scoop.eu

## KEY TERMS AND DEFINITIONS

**Collaborative Business Models:** Emerging Collaboration business models between Original Equipment Manufacturers (OEMs) and Manufacturing Organizations where the OEMs supply machine' on- lease, monitor and maintain using service contracts. OEMs transform from machine sellers to sellers of machine services and Manufacturing Organizations become consumers of machine services rather than machine owners.

**Command and Control Operations Centre:** A command and control Operations center, also known as a situation room, centralizes the monitoring, control, and command of a manufacturing organization' overall operations.

**Condition Based Monitoring:** Condition Based Monitoring is a form of maintenance that involves sensors to measure the real-time performance of an asset while the asset is operating, The data gathered is analyzed to monitor the asset performance, identity patterns/ trends and predict probability of failures based on asset' current state.

**Industrial Automation:** Industrial automation is the use of advanced information technologies to manage processes and machines in an industry to improve operational efficiency, productivity, flexibility and safety.

**Industrial Internet of Things:** Industrial Internet of Things is an emerging technology aimed at embedding devices such as sensors, actuators and other similar instrumentation in machines to create a world of Smart Machines.

**Industrie 4.0:** Industrie 4.0 is the the fourth paradigm shift in manufacturing industry, in which intelligence is built into manufacturing ushering in an era of Smart Industries. Industrie 4.0 is really a "digital" revolution that uses high degree of digital technologies and automation to create a vision for the future of manufacturing.

**Intelligent Optimized Machine:** Intelligent Optimized machine is a machine that operates at peak performance and enables both operating and maintenance costs to be minimized.

**Internet of Things:** Internet of Things is an emerging technology aimed at inter-connecting disparate objects to interchange data - it comprises of sensors and actuators, hardware and firmware electronics, system and application software, and finally the connectivity which "enables" objects to link together and interchange data.

**Smile Curve Manufacturing Theory:** Smile Curve Manufacturing Theory suggests that processes at the end of production value chain i.e. Product Research and Design and Product Marketing and Support functions drive higher financial value as compared to the manufacturing function that supposedly creates least value.

# Chapter 3
# Machine Learning Techniques to Mitigate Security Attacks in IoT

**Kavi Priya S.**

 https://orcid.org/0000-0002-1292-9728
*Mepco Schlnek Engineering College, India*

**Vignesh Saravanan K.**
*Ramco Institute of Technology, Rajapalayam, India*

**Vijayalakshmi K.**
*Ramco Institute of Technology, Rajapalayam, India*

## ABSTRACT

*Evolving technologies involve numerous IoT-enabled smart devices that are connected 24-7 to the internet. Existing surveys propose there are 6 billion devices on the internet and it will increase to 20 billion devices within a few years. Energy conservation, capacity, and computational speed plays an essential part in these smart devices, and they are vulnerable to a wide range of security attack challenges. Major concerns still lurk around the IoT ecosystem due to security threats. Major IoT security concerns are Denial of service(DoS), Sensitive Data Exposure, Unauthorized Device Access, etc. The main motivation of this chapter is to brief all the security issues existing in the internet of things (IoT) along with an analysis of the privacy issues. The chapter mainly focuses on the security loopholes arising from the information exchange technologies used in internet of things and discusses IoT security solutions based on machine learning techniques including supervised learning, unsupervised learning, and reinforcement learning.*

DOI: 10.4018/978-1-7998-0373-7.ch003

## INTRODUCTION

Today's Internet becomes the connectivity of many smart devices and computers. Any real world object can be attached with a sensor and connected to the network. It paves way for many applications that benefits the users. Some common applications are automation in industry, smart home, patient's effective health monitoring applications etc. Some years back, the devices are connected in a network, which is now getting evolved smarter by the connection of any real-world objects. Clearly it states that Internet of Things(IoT) is a fast-evolving technology. Some statistics on IoT predicts that there will be more than 5 billion IoT devices connected at present. IoT can be any physical device equipped with sensors are connected with a communication channel. Through the connected network the devices can interact with the environment, i.e. collect data from surroundings and send that data for processing. Such devices that interacts with the environment to collect data is called as source node. The data is collected by source node and communicated to the base station or the sink node for processing or storage.

Consequently, an algorithm is the responsible for the data collection or data gathering and routing the data to the base station. All these devices are interconnected to share and exchange the data, that makes the IoT and wireless sensor network open to many challenges in security violations and privacy exploration for the users.

## MAIN FOCUS OF THE CHAPTER

In this chapter, provides an idea about the wireless sensor network and IoT, which is an interconnection of the devices controlled through the Human machine interface (HMI). The essential features and use of connected devices or the embedded devices with the network provide a number of uses in many applications. This attractive feature also enables IoT devices connected with the network more prone to security threats and attacks. Depending on the data being communicated over the network, it inhibits an interest over the attackers with a wide range of privacy exploration. Hence providing a secured connected network has to ensure it provides solutions for the various concerns like Privacy of data, data reliability, correct responses from the connected devices, trust-worthy devices and autonomous recovery of the device when compromised. Considering these factors, the IoT requires effective solutions to achieve the above terms.

# TECHNOLOGIES CONNECTING VARIOUS IOT DEVICES

The main objective of the Internet of Things is to provide an environment in which the connected devices are able to transfer information without any manual interference. Thus, the exchange of information between two devices is possible under some well-established communication technologies, which are discussed below.

## Wireless Sensor Networks (WSN)

Wireless Sensor Networks are comprised of set of independent nodes with limited bandwidth and frequency through which it can communication wirelessly with other nearby devices. In traditional wireless sensor network environment, the sensor node consists of the following parts:

1.  Sensor
2.  Microcontroller
3.  Memory
4.  Radio Transceiver
5.  Battery

The sensor nodes in the wireless sensor network has very limited communication range (short range communication). Hence the communication becomes multi-hop relay of information between the source and the base station. The required data is collected by the wireless sensors through collaboration amongst the various nodes, which is then sent to the sink node through a suitable routing strategy. The communication network formed dynamically by the use of wireless radio transceivers and it facilitates data transmission between nodes. Multi-hop transmission of data demands different nodes to take diverse traffic loads.

## Radio Frequency Identification (RFID)

In context to the Internet of Things (IoT), RFID technology is mainly used in information tags interacting with each other automatically. For exchanging information between one another and interaction between them the radio frequency waves are used. There are some components being used in this RFID technology. The major two components are:

## RFID Tags

RFID tag is a small device in which a small chip is embedded with an antenna. It serves as a unique identifier with a help of a code known as Electronic Product Code (EPC) which is also stored in the memory unit of the RFID tag. EPC provides a feature to recognize the particular tag universally with some numerical form of data associated with it. A tag reader is allied with the RFID tag to operate on the universal tag code The tags are of two main types as:

1.  **Passive Tag:** These tags are activated only by a transceiver from a specified range of distance which relays on the information of the EPC.
2.  **Active Tag:** The distinct EPC interacts with all its contiguous EPCs available distantly at some limited range. Active tag has an internal battery associated with it to facilitate this behavior.

## RFID Readers

The RFID reader functions as the identification detector of each tag by its interaction with the EPC of the tag under its scan.

# SECURITY AND PRIVACY CATEGORIES AND PROBLEMS

IoT is the network of devices embedded in motor vehicles, streets, apartments or buildings and other electronic appliances that enables the devices to collect and exchange the information. This valuable information or the service provided can be breached and compromised by attackers or illegitimate users in the motivation of financial gain or damage the reputation of the competitor. Some of the common attacks are:

1.  **Device Cloning:** In device cloning security issue, any foreign/alien device can be connected to the network as an authenticated device, but it is not. It is even harder to find which are the clones and which are authenticated devices. The foreign device will quickly overload the server with bad data, causing massive breakdown leading to time delay and heavy financial loss to fix it.
2.  **Exposure of Sensitive Data:** Information provided across the network needs to be encrypted. When the sensitive information is not properly protected by the application through standardized protocol or encryption technique then

Sensitive Data Exposure occurs. A simple example can be transferring the data as clear text which is rendered as ASCII and can be read by text editor.

3. **Denial of Service:** In simple terms this is denying or slowing down the service. The attack on the network causing the network to slow down or even completely pull the network down by simulating the network with unwanted useless data creating a heavy traffic in the data flow. Variations of DoS threats are directly targeting the server infrastructure while others exploit the vulnerabilities in applications.

4. **Unauthorized Device Access:** There are concerns about applications that are too private which does not digest third party to control the devices. A malicious user may gain control of the devices over the network.

IoT extends it applications in almost all aspects of modern life. Therefore, security is a very crucial factor. Furthermore, the limitation of the capacity and energy of the device complicates this problem.

Security in IoT is a problem that cannot be easily solved. One major factor is that the IoT devices are battery operated and with limited resources. This also makes the IoT device very inexpensive. The security techniques consuming more energy and memory resources are not a feasible solution. Another factor is the IoT devices are adapted to dynamic changes, so it is not suitable to apply a centralized security algorithm. Machine learning techniques can be one effective solution.

This chapter elaborates the different security threats that are imposed on IoT possible machine learning techniques that can be developed to encounter these security threats.

## Classification of Security Attacks in IoT

The most critical attacks in IoT can be classified as different security attacks such as Attacks on the layers, attacks based on the targets and attacks based on the performer.

- **Attacks on the Layers:** These attacks are focused on the different layers such as Application layer, Transport layer, Physical layer, Network layer and Data-link layer. Each of these layers is open to different types of attacks. For instance, flooding of the data packets in the data link layer in which nodes will suffer from high collision of data. Another instance, where the malicious nodes sends more requests to transmit the data and so deplete the energy in the batteries of other nodes causing exhaustion attacks leads to failure of the network.

- **Attacks Based on Targets:** These types of attacks focused on the targets are performed to threaten the target with data confidentiality. In these attacks, the unauthorized or malicious attackers are exposed with the information of high confidentiality such as the keys without the knowledge of the authorized users. The attackers then decrypt the keys that are weak and gain the information needed. Traffic analysis and eavesdropping are some examples of passive attacks. Meanwhile, active attackers may obtain confidential information through monitoring the network and may even change this information. Some examples of active Attacks are hole attacks, spoofing, Sybil attacks etc.,
- **Attacks Based on the Performer:** Based on the attacker's location in the communication network, we can classify as outside attacks (outside the network) and inside attacks (from inside the network). In case of the attacks performed from inside the network, one of the legitimate or authorized nodes will be the attacker. Hence, these authorized nodes can access any information required and it cannot be easily identified. These attacks can lead to data modification and eavesdropping of the data. In case of outside attacks, the attackers may send unwanted data again and again and can increase the network traffic with unwanted data. When the network is over jammed with data, the resources may lose all the energy and becomes exhausted.

Though the IoT is widely implemented in various applications, the entire communication network and substructure and setup of the IoT is deviated from the standards creating a flaw on the network. The IoT network is always suspected with confidentiality loss for the users of the IoT applications. The overall development of IoT system is affected by some predominant security threats varies over the network since this technology is built over the communication of the information relay on one or more devices. Some of the most prominent security issues arising out from the communication technology are the following:

## SECURITY ATTACKS ON SENSOR NETWORKS AND IOT DEVICES

Security attacks in sensor networks can be broadly classified into two broad classes: passive and active attacks. In passive attacks the attackers focus mainly against data confidentiality, But in case of active attacks, malicious attacks are performed against data confidentiality and also against data integrity. Active attacks can also aim for the resource utilization or the disturbance of any communication and unauthorized access. An active attack can be detected, but passive attacks are a bit hard to detect.

## PASSIVE ATTACKS

In passive attacks the malicious nodes are not known and remain hidden to tap the communication channels to collect data. It can be defined as the monitoring and listening of the data flow channel by unauthorized intruders. The passive attacks are classified into traffic analysis types, packet-tracing, eavesdropping and camouflaged adversaries. The categorization of passive attacks is shown in Figure 1. Some other types of passive attacks include camouflage and Packet-tracing.

### Monitoring and Eavesdropping

The data classified in the communication network can be monitored by tapping over the communication channel known as eavesdropping. Compared to the wired, wireless links are easier to tap. Hence, the wireless communication networks are more susceptible to passive attacks. In particular, when known security standards are used and plain data, i.e. not encrypted, are sent wirelessly, any compromised intermediate node can easily receive and read the data and listen to or watch audio–visual transmissions. For example, an adversary can easily eavesdrop credit card numbers and passwords when they are transmitted without any encryption standards over unsecured wireless sensor networks.

### Camouflaged Adversaries

In this type of passive attack, the intruder can hide the number of nodes available in the wireless sensor network. So such nodes will show themselves as the legitimate nodes and intimate the other nodes to send the data packets. On receiving the packets, these nodes will misroute the packets or can perform a detail analysis on the data

*Figure 1. Major types of passive attacks*

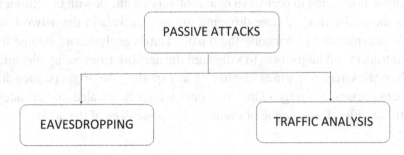

being private. The term camouflage is used because in this attack the sensor node compromises the other nodes in network and it also makes a false advertisement about the routing information such that all the data packets flows through this camouflaged node. Once the packets are received the data is forwarded and analysis is done on the private data being forwarded.

## Packet-Tracing

In this packet tracing attack, the malicious intruder may themselves notify the position of this node and immediate sender of the other packets originating. The malicious node is capable and equipped in finding and tracking hop by hop neighbors thus leading to the actual source node from which the data is originated. Thus, the main source of the data is revealed to the malicious node causing the exploration of the privacy of the data.

## Traffic Analysis

The traffic pattern of the flow of data packets in the communicating network has equal importance as the data present in the packets, and provides valuable information for the intruders. For example, by analyzing traffic patterns of the communicating network, the topology of the network can be explored. In case of wireless sensor networks, the nodes deployed at random places transmits data to the neighbor node that are very close to the base station. This node nearer to the base station is called sink node. Ultimately, the sink node makes more data transmissions compared to the other nodes, because the sink node closer to base station is responsible for many data transmissions.

Similarly, in ad hoc networks, to ensure the scalability factor clustering is an important factor. In each cluster there will be a cluster head which will always be responsible for the data transfer from other nodes in the communicating network. It is very useful for the attackers if the cluster head is compromised or the base station itself is detected to perform a denial-of-service attack with the cluster head denying the traffic flow or eavesdropping the data packets in the network to get valuable information by analyzing the traffic. Traffic analysis can also be useful for the intruders and helps them to safeguard the network from anonymity attacks. The adversaries may also aim at identifying and spotting the origin (source nodes) of the data packets as a target. Once this information is revealed, the intruder can detect the abilities, flaws, scene of events of the possessors of the nodes.

The classification on the security threats and issues faced in WSN is shown in Figure 2. In addition, there can be many other security attacks are possible in a wireless sensor network falls under three distinct classification listed below:

1.   Security violation on authorization.
2.   Breakdown of the integrity in privacy through silent outbreaks.
3.   Security breach on the network availability.

## ACTIVE ATTACKS

In active attacks the main intention of the malicious node or attackers is to affect the data flow in the data communication network. With the sign of this objective there is a possibility that the attacker and can be detected. For example, there may be degradation overall data flow and network traffic flow because of these attacks. In most cases the intruders will try to stay hidden and aims to gain unauthorized access to the network. The adversaries can also perform this attack to disturb integrity of the network and also can be a threat against confidentiality. The active attacks are classified into four groups. Figure 3 depicts the classification.

*Figure 2. Hierarchical classification of security issues*

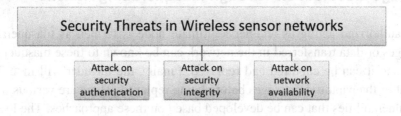

*Figure 3. Types of active attacks*

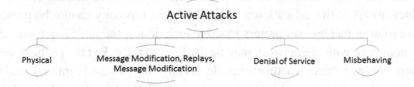

## Physical Attacks

Another possible attack is the physical damage of the hardware by the intruder or system to terminate the nodes from the network and can even terminate the communication network by damaging half of the nodes available in the network. The physical attack is also considered to be responsible for affecting the fault tolerance of the network, i.e. the capability to withstand the node failures and provide all the functionalities of the network. Mostly in wireless sensor networks, the deployment of the sensor nodes is random and installed in unattended regions, which can be accessible by the intruders. Hence, they can be physically damaged by the attackers or easily defected and push out of the network. When nodes are deployed and unattended it can also be reached physically and the node can be tampered with techniques like micro-probing, laser damage on the nodes, glitch attacks and power analysis. The act of tampering the nodes can as well lead to the DoS attacks and masquerading attacks.

Therefore, the node should be resilience to tampering attacks and thus it is an open area that sensor network and its promising applications should seriously consider.

The node-tampering schemes can be grouped into invasive tampering and non-invasive tampering. Gaining unlimited access to a node is called Invasive tampering i.e. this is intentionally done to access the node. In contrast gaining such a boundless entrée to the node is not the aim of non-invasive attacks. Instead, the intruder performs this attack to analyze the properties of a node, such as the memory capability, battery energy consumption, or the time complexity and space complexity of the algorithms implemented.

## Message Modification Through Masquerading Attacks

An unauthorized acting as another legitimate node is termed as masquerading. Messages or data transferred in the network can be known to these masquerading nodes and it can be captured and replayed. Finally, the intruder will modify the content of the captured messages before being replayed. There are various threats and vulnerabilities that can be developed based on these approaches. The location of the nodes in the network may dynamically change due to its movement in mobile ad hoc networks. So, there is no fixed location for the sensor nodes. The nodes inhibit autonomous mechanisms like auto-forming and auto-healing to adjust to the fluctuations in the network topology. Since the topology cannot be preserved and the routing mechanism prefers reactive techniques, the node's consistency in correspondence with the network may be difficult to trace. But the problem here is it is impossible to detect whether the node has any other access point in the existing network. The sensor networks do not use the global identifications that makes the network even easier to masquerade the network.

The data integrity of messages or the service being provided in the network can be attack through any type of attacks and may lead to modification of data violating the integrity. The act of message modification and replay of messages through masquerading attacks is also a security threat against data confidentiality. It creates an illusion that the node is a legitimate node and make the nearby nodes to show the trustworthy data to itself or another wicked node to access the private data. An intruder deceives someone and make the victim to give confidential information voluntarily through phishing. An unauthorized node that pretends as if it is an authorized node. It requests passwords, encrypted keys and other information from any other node. One such approach is the act of Masquerading that attacks and access the system illegally.

## DOS ATTACK ON THE PHYSICAL LAYER

In a wireless sensor network, the physical layer is responsible for modulation and demodulation, encryption and decryption process. Since this layer provides the much needed functionality, it is highly prone to more security attacks. This security attacks that can be performed on this are mentioned below:

1.  **Network Jam:** The communication network of the sensor nodes is compromised and jammed. This network causes the sensor nodes unable to communicate with other sensor nodes.
2.  **Node Tampering:** The motive of this attack is to physically tamper the particular sensor node in order to extract some sensitive or confidential information.

## DOS ATTACK ON THE LINK LAYER

The data streams generated in the network, detection of the data frames are processed in the link layer of wireless sensor networks. The MAC and error control mechanism is also responsibly processed by the link layer. Moreover, the reliability of the data from one point to another point or between multiple points are ensured by the link layer. The various attacks competing in this layer are:

1.  **Collision of Data:** When two or more nodes transmits the data packets simultaneously on the communication channel, there will be a delay in the processing which leads to this type of attack. Due to the collision of data

packets there may be minor changes in the packets. When the data packets are received at the other end, due to the minor changes inhibited it is identified as a packet mismatch.

2. **Unfairness of Data:** It is a vigorous form of the repeated data collision attack, which in turn creates a major change to the data packets making it unfair at receiving end.

3. **Exhaustion of Battery Life:** This attack causes unusual loss of the energy in the nodes due to unusual high traffic in a channel, depleting all the energy of the sensor nodes. The traffic of the channel may be increase due to producing large number of requests (Request to Send) and increased number of responses over the channel.

## DOS ATTACK ON THE NETWORK LAYER

The network layer is responsible for the routing of the data traffic in communicating network. Hence the network layer is open to many attacks to compromise the routing strategy. The DoS attack is the more predominant attack in this layer. The specific attacks are:

- **Hello Flood Attack:** A single malicious node sends a useless message (Hello message), which flooded to all the nearest neighbor nodes. These neighbor nodes will then forward and replay the messages to other nodes creating a high traffic and congestion in the communicating channels. At one point the complete network is flooded with these unwanted packets and congested.

- **Homing:** This kind of attack is like focusing on the root of the processing. In homing attack, the capability of the network is compromised by searching and targeting the cluster heads in the network. Using these host nodes, the entire network can be shut down.

- **Selective Forwarding:** Normally the data is forwarded to the nodes that needs the packets. In this selective forwarding attack, the data is transmitted to a selective number of nodes by the compromised node. The compromised node may select the nodes to which they should forward the data, based on the requirement and objective of the malicious attack.

- **Sybil Attack:** A single node in the network is presented as multiple number of identities. So many nodes forward the data to this node.

- **Spoofing:** A special kind of attack which plays on replaying and misdirection of traffic.
- **Wormhole Attack:** The data packets are relocated from its original position.
- **Sinkhole:** It is a special kind of attack in which the nodes are convinced to forward the packets through the malicious node. The unauthorized node pretends as the sink node and gets all the data packets or the malicious node convinces the surrounding nodes that it is nearer to the sink and attracts the data packets. When this malicious node pretends to be the sink node, it becomes the centre and receives all data packets from the neighbors. This also paves way for many other attacks like wormhole attacks, and tamper the data by selective forwarding attack.
- **Black Hole Attack:** The illegitimate node pretends to be the neighboring node of the sink node. The power transmission of these nodes will be higher and all other nodes carry data to this node making the network vulnerable. All the data packets received will not be forwarded but dropped off by this malicious node. At some point of time the entire network traffic flow will be stopped creating a black hole region.
- **Acknowledgement Flooding:** In routing algorithms to ensure the data being received acknowledgements are used. In this flooding attack, the neighboring nodes are spoofed with a false information and forwards the acknowledgments to the destined node.

## DOS ATTACK ON THE TRANSPORT LAYER

The transport layer of the wireless sensor network architecture makes available the functionality and consistency of the data communication. The DoS attacks in this layer are:

- **Flooding:** It refers to deliberate congestion of communication channels through relay of unnecessary messages and high traffic.
- **De-Synchronization:** In de-synchronization attack, fake messages are created at one node or at both endpoints nodes requesting retransmissions and corrections of non-existing error. This results in loss of energy in one or both the end-points in carrying out the retransmissions.

# DOS ATTACK ON THE APPLICATION LAYER

The application layer of the wireless sensor network conveys out the accountability of traffic organization. These applications bring out the transformation of data into an understandable form or supports in gathering of information by sending enquiries. In this layer, a path-based DoS attack is introduced by stimulating the sensor nodes to generate an enormous traffic flow in the path headed for the base station. Some additional DoS attacks are as follows:

1.  Greedy Attack
2.  Interrogation
3.  Black Holes
4.  Node Subversion
5.  Malfunction of node
6.  Passive Information Gathering
7.  False Node
8.  Message Corruption

# SECURITY CONCERNS AND THREATS IN RFID TECHNOLOGY

In the area of IoT applications, RFID technology plays a major role as the RFID tags are used for autonomous information exchange without any human intervention. The use of RFID tags paves way for more risks and open to security threats since it is open to outside attacks due to less security feature in the RFID technology. There are many security attacks and issues, out of which most common types are listed below:

1.  Unauthorized authenticity: The RFID tags losses its capacity temporarily or permanently because of the DoS attacks. The RFID tag will start to malfunction and misbehave in scanning the tag reader when it is being compromised or attacked.
2.  The attacker can perform these DoS attacks remotely, allowing them to manipulate the RFID's behavior.
3.  Unauthorized cloning: The RFID tags can be manipulated through which the confidential information can be captured falls in this category. The cloning or replication of the tag is possible, once the tag is compromised thereby introducing new vulnerabilities.

4.  Unauthorized tracking: An adversary traces the RFID tag's behavior which can result in providing privacy or confidential data like a person's phone number etc.

5.  Unauthorized Replay: The signal used for communication between the tag and the reader is intercepted and known. This message or signal can be replayed at a later point of time by modifying the message, thus providing a fake availability of the RFID tag.

6.  In addition to these threats some additional security issues in RFID technologies are:
    ◦   Middle man attack
    ◦   Tracking
    ◦   Reverse Engineering
    ◦   Viruses
    ◦   Eavesdropping
    ◦   Killing Tag Approach

It is also very important to take into consider on social engineering (i.e.) the human responsible for the IoT security framework also plays an important role for the management of the confidential data and are responsible for the rules to be ensured for maintaining the security. Some of the security management and security rules are:

•   New security constraints, instructions and interactions are introduced.
•   The efficiency of the rules is examined.
•   The rules are set into operational mode and practiced.

In interconnected networks where two or more systems are connected, there can be many deviations in the security standards. These deviations will lead us into the real-world security problems. The security concerns also need to be extended to protect the individual data such as financial exchanges, person's health data, etc. In critical real-world applications that function with control systems, such as automatic control of vehicles(cars) and nuclear reactor, will lead to more serious damage of human life when compromised with attackers. Some of the major problems highlighted below:

•   Protecting unauthorized interference of data.
•   Preventing the endpoint devices from the unauthorized control.
•   The increase in the development of the network evolves the threat on Cyber security.
•   Updating security capabilities of IoT devices post installment.

## MACHINE LEARNING SECURITY IN IOT

Machine learning (ML) techniques plays a quite wide role in security attacks and IoT devices should be able to choose a good defense policy against these smart attacks. The heterogeneity and dynamic behavior of the network makes the security protocols vulnerable and yet the key parameters should be determined. This becomes a more challenging task since the IoT device is equipped with very limited resources, which makes the device difficult to estimate the accuracy and providing efficient counter attack in the communicating network.

For example, the required information like authentication performance and sensitive test threshold values used in the hypothesis test for the applications with outdoor sensor deployment leads to a higher false alarm rate. To overcome such problems and to improve the efficient performance of the IoT system, one possible solution is to couple the system with machine learning techniques for improving security considerations. To develop the network safety such as detection of malwares, unauthorized access, anti-jamming offloading and illegal access control, we can apply machine learning techniques like supervised learning, unsupervised learning, and reinforcement learning.

**Supervised Learning**: The learning is termed as supervised, if the inputs are clearly known with their desired outputs. The dataset is provided with each input associated with an output. This training dataset is provided as the input to the machine. The machine can identify the inputs and its corresponding outputs. The algorithms under supervised learning are: neural networks, deep neural network (DNN), K-nearest Neighbor(k-NN) and random forest algorithm. These algorithms are applied in the IoT devices to build the required classification model to label the network traffic. In addition, we can also couple the security enhancement with Support Vector Machine (SVM).

Some instances where these learning algorithms applied in IoT devices are:

- Naive Bayes and SVM may be developed by IoT devices in network invasion discovery and network spoofing.
- DNN can be applied in IoT devices with sufficient memory resources and computation to detect spoofing attacks.
- To spot the network invasion we can use K-NN and random forest classifier for malware discoveries.
- Employ neural network to sense DoS attacks.

**Unsupervised Learning**: In contrast to the supervised learning technique, the unsupervised learning is provided with the input data only. Considering these inputs, the model will classify the data into different groups called clusters. Unlike the supervised learning that require labeled data, unsupervised learning takes only the input and finds the similarity between the input data. Using the similarity measures, unsupervised learning clusters them into different groups. Multivariate correlation analysis is done by the IoT device and can apply unsupervised learning to ensure the privacy of the network.

**Reinforcement Learning**: This type of learning is based on the reward scheme. No proper inputs are provided into the system. The learning is based on the previous positive outputs. Examples are playing cricket, playing a chess game, in which based on the previous move, we need to decide the next move. If the performance is good, for instance, if the opponent player's chess coin is eliminated, then the move is effective and it is rewarded and if not, no reward is given. Such type of learning is called Reinforcement learning. The learning also depends on the surrounding environment. One such example of reinforcement learning is Q-learning. IoT device can be enabled with some RL techniques such as Q-learning with different parameters to defend against various attacks by choosing the safety practices. Such type of learning ensures the privacy authentication and to detect malware and helps in anti-jamming transmissions.

In this chapters, the major considerations are malware detections in IoT, unauthorized access control and ML-based authentication to provide secure offloading.

## SECURITY CONCERNED SOLUTIONS THROUGH MACHINE LEARNING

In real time IoT device implementation, there are many challenges and security considerations that are to be estimated properly, since they may lead us to security threats and issues.

1.  Authentication of IoT Devices is more important and the device should authenticate itself. Only after proper authentication it should starts its transmission of data and receiving data.
2.  Firewalling to ensure and allows the secured and trusted use of the packets. The devices will communicate with one another after implementing the algorithm through this firewall that provides secured communication.

3. Access control, to limit the control and privileges to the device components.
4. End-To-End Encryption, which is equipping the devices with security encryptions by implementing the software on all derives. Also preventing the interference of the unauthorized access so as to reduce security threats.

## LEARNING-BASED AUTHENTICATION

With inadequate computational, memory and energy possessions it is not always applicable for IoT devices to incorporate traditional authentication schemes to distinguish identity-based attacks. In Physical (PHY)-layer verification technique, the MAC address provides security for the privacy information with a light-weight security fortification for the IoT devices. The parameters of the transmitters and radio receivers such as received signal strength (RSS) and he received signal strength indicators (RSSIs) are taken into consideration. This technique also considers the feature exploited by the PHY-layer such as channel state information (CSI) and provides security with less computation and message overheads. Nevertheless, it is thought-provoking for an IoT device working in a diverse environment with unstructured communication network, to indicate a suitable trial threshold of the endorsement and the mysterious spoofing or attacking model. The simple solution is the IoT devices can relate any Machine Learning(ML) techniques. Some ML technique-based authentication can be enabled in the IoT devices to maximize the accuracy of the authentication and to improve the utility by achieving the optimal test threshold. This technique is very similar as a Markov verdict procedure (MVP), in which the key is to determine the authentication parameters and can be made as an IoT authentication game without the conscious of the networking prototype.

The IoT system can adopt the incremental aggregated gradient (IAG) and the Frank-Wolfe (dFW) technique to improve the spoofing resistance. Such supervised learning techniques can improve the spoofing detection accuracy and also reduces the overall communication overhead. For instance, the IoT device is implemented with the authentication scheme with dFW and IAG technique to prevent spoofing attacks.

An unsupervised learning is used to authenticate the IoT devices. A non-parametric Bayesian method under unsupervised learning is used for the authentication scheme for the identification of the IoT devices. To evaluate the arrival time of the data packets, RSSIs and to monitor the intervals of the radio signals, the IGMM proximity scheme is applied in the IoT device to authentication the device. The information about the location of the devices is not revealed easily in this scheme. Through evaluating these parameters to detect malicious intruders and attackers outside the proximity range. This scheme will request the IoT device to send the ambient signals and packet arrival time internal during a specific time duration. The legal legitimate receiver in-turn will receive the data signals from the IoT device. After such authentication messages are received, the receiver applies a suitable machine learning algorithm to detect and compare the signals reported. The authenticated information is provided to the IoT resources by the IoT devices. Finally, the IoT devices are applied with the deep learning techniques such as DNN with the available computation to improve the authentication accuracy.

## LEARNING-BASED ACCESS CONTROL

Many different kinds of nodes are made available in heterogeneous networks and source of data will be diverse that leads to a challenging design to access control. Machine learning techniques such as neural networks, K-NN, SVM, can be used to detect the malicious intrusion. For instance, multivariate correlation analysis (Liang Xiao, 2018) can be used to detect the DoS attack. By this scheme the network traffic features and their correlations are extracted. The accuracy of this scheme in detection increases by 2.95% to 94.8% related with the nearest neighbor-based methodology through triangle area-based approach. The IoT devices such as outdoor sensor normally has limited resource and computation constraints, that makes the device with degraded performance in intrusion detection techniques. Machine learning techniques help the IoT devices to build an effective access control protocols by ensuring extended network lifetime and conserved energy. The outlier detection scheme (Xiaoyue Wan & Xiaozhen Lu, 2018) is developed to improve the flexibility with reduced energy consumption and also to address the outlier detection in WSNs by applying the K-NN learning technique. The traditional centralized scheme when compared with the above learning technique provides an increase of the energy conservation by 61.4%. K-NN is a modest machine learning technique in which the inputs are classified with the K nearest neighbors. Euclidean distance measurement can be used to find the distance measurements between the IoT devices.

The technique proposed on detection of the unauthorized access control (Yanyong Zhang & Di Wu, 2018), utilizes a multilayer perceptron (MLP) with a hidden layer that consists of two neurons. The neural network is trained with the input connection weights and the identifying factor is introduced. Thus the MLP is trained to identify if there is a possibility of DoS attack on the device. The backpropagation (BP) that forwards the calculated results and back propagates the error rate that should be implemented. An evolutionary computation technique like particle swarm optimization can also be implemented that uses the particles to modify the associated connection weights of the perceptron. The IoT device implemented and tested in such scheme proves that the network lifetime is extended by improving the conservation of the device's energy, if the MLP exceeds a desired threshold.

## LEARNING-BASED IOT MALWARE DETECTION

To detect the malware, the IoT device can evaluate the runtime behaviors of network by applying the supervised learning techniques. In the malware detection scheme, random forest classifiers and K-NN can be used by the IoT device to develop the malware detection model. The TCP packets are filtered in the IoT device. Further on considering the selection of the feature, this scheme includes the various network features like the length of the frame and frame number. These features are labeled and then selected. In the K-NN based malware discovery the data movement in the network is assigned with many sensor nodes or devices among the K closest neighbor nodes. To distinguish malwares, the network traffic labeled with the decision trees is built with random forest classifier. The analysis on the experimental results shows that the random forest based experimental results and K-NN based malware results are almost very close with true positive rates around 99.7% and 99.9%, respectively on the MalGenome dataset.

### K Nearest Neighbor algorithm

A modest machine learning algorithm is classifying the inputs based on the adjacent neighbors. The network is comprised of number of nodes and each node has a set of neighboring nodes. This is the K-NN algorithm which is a lazy learning method with a set of testing phase only, and no training phase is available. So the typical K-NN algorithm has to dependent only on the test instances. The K-NN algorithm is provided with the entire dataset and stored. The algorithm looks up for the matching pattern, so it does not require any extra training phase. The erroneous data are removed from the database so as to maintain consistency. When a new instance arrives, predictions are made by recognizing the most alike neighbor and defines the output class label. The estimation of the similarity measures can be calculated with any distance-based measures. Euclidean distance, Manhattan distance are some of the distance measures that can be used.

## Random Forest

In the training phase huge quantity of decision trees is built using the Random forest techniques and it is used to find the class label. Decision tree comprise of a tree like graph to sort conclusions. The nodes at the leaves in the decision tree are used to represent the attributes. Also, the nodes and edges can be used to represent the conditional attribute values. The path from the root to the leaf node expresses the grouping rubrics. A sample decision tree with binary classification is picturized in figure 4.

The nodes A1 and A2 signify the characteristics, and the conditional values are referred in the edges for finding the class labels. If the value of the attribute A1 is ≥ 0.5, then it is classified as class1. If it is less than the desired value, then it flows to the next lower level value and checks the value of the attribute A2. This explains the working of the decision tree. This enhances the performance of the random forest model. The set of input class labels and the associated set of output class labels is given as input to the algorithm. This is training phase, where the algorithm is trained with a set of inputs associated with its corresponding outputs. Once the training phase is done, the results obtained from the input classification are considered for the testing the new input data. This training process with many classified input data is repeated so as to improve efficiency of the algorithm to classify new data.

*Figure 4. Sample decision tree classification*

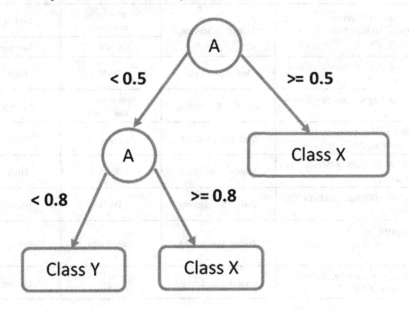

Finally, any new input dataset is provided and predictions are done with the trained algorithm. The IoT devices should be capable of detecting the malwares traces in the security servers provided in the cloud or the edge devices. Further the security services should be more powerful with high computation speed and should store bulk amount of data with larger malware database.

Without the aware of the model, the IoT device is applied with the Reinforcement Learning (RL) techniques to achieve the optimal offload. To increase the malware detection rate, the offloading rate can be improved by applying the Q-Learning in the IoT device. This scheme enables the IoT device to estimate with increased accuracy gain and better detection ratio.

Table 1 provides the summary of various approaches in machine learning for secured IoT and WSNs. Though different approaches yield high accuracy, the implementation of the support vector machine is more complex than supervised learning techniques. There are quiet many encounters in acquiring the security of IoT and WSNs since machine learning should compromise between a dense layer of security and a low computational complication to be appropriate for the limited resources.

*Table 1. Summarizes the various ML techniques to ensure the security*

| Reference papers | Type of attacks | ML techniques | Complexity |
|---|---|---|---|
| Enhancement of Security (Kaplantzis, Shilton, Mani, & Sekercioglu, 2007) | Denial of Service | Support vector machine | Simple |
| | | DNN | Average |
| Distributed attack detection (Alajmi & Elleithy, 2015) | Denial of Service | DNN | High |
| Data forwarding attacks (Ferdowsi & Saad, 2017)[ | Selective Forwarding | Support vector machine | Average |
| Machine Learning to secure IoT (Miettinen et al., 2017) | Middle man attack | Supervised | High |
| Dynamic Watermarking (Meiden, et al., 2017a) | Middle man attack | RNN | High |
| IoT machines (Rathore, Badarla, Jha, & Gupta, 2014) | Traffic Monitoring | DNN | Average |
| Compromised Devices (Outchakoucht, Hamza & Leroy, 2017) | Traffic Monitoring | Supervised | Average |
| Bio-Inspiration (Rathore, Badarla, Jha, & Gupta, 2014) | Unauthorized node | Support vector machine | Average |

# EXPERIMENTAL ANALYSIS ON INTRUSION DETECTION

This section deals with the different experimental analysis results of various techniques. Different experimental analysis is done to estimate the performance measures and accuracy rate.

The dataset selected for analysis is the standard KDDCUP99 dataset for intrusion detection. This dataset has many redundant records. This redundancy of data possibly will lead to biased results. So the NSL-KDD dataset that is an advanced version of the KDDCUP99 dataset is selected and the supervised learning algorithms are tested on that dataset. It has a total number of 42 features and the four diverse categories of simulated assaults to experiment the invasion detection system.

- **Denial of Service (DoS) Attack:** Non-availability of the system resources and heavy usage of the bandwidth leads to the DoS attacks.
- **User to Root (U2R) Attack:** At initial point the invader has access to the normal user account, then slowly gains admittance to the root system by abusing the resistances of the system.
- **Probe Attack:** Before performing or introducing any attack having an access to the entire network.
- **Root to Local (R2L) Attack:** By ill-using some of the liabilities of the network attacker achieves access to local devices by transferring packets on a remote device.

The data collected is initially preprocessed and divided into testing phase data and training phase data. The data available in textual format are completely converted into numerical data. After conversion of these data into numerical form, then the models to work on that data are built using any of the learning algorithm such as Support Vector Machine, Logistic Regression, and Random Forest classifiers. The test data provided are predicted by any of these learning algorithms implemented. The labels predicted by the models are compared with the actual labels. The methodology and experimental analysis are built with the following steps:

1. The cleansing of the dataset – preprocessing step.
2. The separation of the training dataset and testing data.
3. The implementation and build of the classifier models for the training data.
4. Predicting the new testing datasets.
5. Testing the algorithm with the training and testing dataset.
6. Comparisons of the accuracy rate of all models.

Table 2 summarizes the performance measure like Precision value, Recall value, and Accuracy rate of the supervised machine learning classifiers in pinpointing the intrusion. The values tabulated are the experimental results from the methodology implemented.

Figure 5 shows the pictorial bar chart to depict the performance analysis of the different machine learning models. Based on the outcomes it can be identified that Random Forest classifier with the peak accuracy, outstrips the other methods. While SVM has the lowermost accuracy, Logistic Regression algorithm has the good accuracy than Gaussian Naive Bayes and Support Vector Machine.

*Table 2. Performance measures of different classification models*

| Classifying Models | Precision value | Recall value | Accuracy rate |
|---|---|---|---|
| LR | 0.825 | 0.848 | 0.843 |
| GNB | 0.786 | 0.809 | 0.791 |
| SVM | 0.764 | 0.789 | 0.746 |
| RFC | 0.988 | 0.991 | 0.987 |

*Figure 5. Performance measures graph*

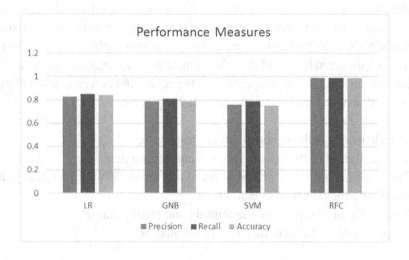

## CONCLUSION AND FUTURE WORK

In this chapter, the different IoT and WSN attacks are identified. The promising challenges faced by the IoT device security in terms of detection of the malware, illegitimate access control, authentication of the IoT devices are discussed. In the consideration of the practical IoT applications numerous challenges should be addressed to implement the learning-based security techniques. In addition, when we implement machine learning techniques the IoT network should withstand the ruthless schemes at the start of the learning (training) process. Generally, it has difficulty in identifying the state of the attack precisely and should evade any security catastrophe at the beginning of the learning process. The protection of the IoT systems also depend on the backup mechanisms. To achieve the optimal strategy, more methods on the Reinforcement learning has to be explored and identify the cause of the network disaster at the learning stage. Some failures on the learning process such as bas feature extraction, oversampling and insufficient training data can be eradicated through supervised and unsupervised learning techniques. Therefore, to offer consistent and protected IoT services, the security solutions have to be incorporated with the machine learning based security schemes. Yet, many surviving Machine learning based security schemes have rigorous calculation and message delivery overheads, and need more training on data classification. And so, new ML methods using little computation and communication overhead are required to enrich security for IoT systems.

## ACKNOWLEDGMENT

First of all, the authors thank the management and the principal of Ramco Institute of Technology and Mepco Schlenk Engineering College for encouraging us to perform such fruitful work and provided us with all the resources we needed. The authors would also like to extend thanks to their family members, colleagues, working partners and beloved friends for their supportive and heartening words in every single phase of this effort. Writers also soulfully be grateful to God, for his elegance and support to make this episode an achievement.

# REFERENCES

Alajmi, N. M., & Elleithy, K. M. (2015). Comparative analysis of selective forwarding attacks over Wireless Sensor Networks. *International Journal of Computers and Applications*, *111*(14).

Ferdowsi, A. & Saad, W. (2017). Deep learning-based dynamic watermarking for secure signal authentication in the internet of things.

Kaplantzis, S., Shilton, A., Mani, N., & Sekercioglu, Y. A. (2007). Detecting selective forwarding attacks in wireless sensor networks using support vector machines. *Proceedings of International Conference on Intelligent Sensors, Sensor Networking and Information*. 10.1109/ISSNIP.2007.4496866

Meidan, Y., Bohadana, M., Shabtai, A., Ochoa, M., Tippenhauer, N. O., Guarnizo, J. D., & Elovici, Y. (2017a). Detection of unauthorized IoT devices using machine learning techniques. Retrieved from https://arxiv.org/abs/1709.04647

Meidan, Y., Bohadana, M., Shabtai, A., Ochoa, M., Tippenhauer, N. O., Guarnizo, J. D., & Elovici, Y. (2017b). ProfilIoT: A machine learning approach for IoT device identification based on network traffic analysis. *Proceedings of the Symposium on Applied Computing*. 10.1145/3019612.3019878

Miettinen, M., Marchal, S., Hafeez, I., Asokan, N., Sadeghi, A. R., & Tarkoma, S. (2017) IoT Sentinel: Automated device-type identification for security enforcement in IoT. *Proceedings of IEEE International Conference on Distributed Computing Systems*, 283.

Outchakoucht, A., Hamza, E. S., & Leroy, J. P. (2017). Dynamic access control policy based on blockchain and machine learning for the internet of things. *International Journal of Advanced Computer Science and Applications*, *8*(7), 417–424. doi:10.14569/IJACSA.2017.080757

Rathore, H., Badarla, V., Jha, S., & Gupta, A. (2014). Novel approach for security in wireless sensor network using bio-inspirations. *Proceedings of International Conference on Communication Systems and Networking*. 10.1109/COMSNETS.2014.6734875

Xiao, L., Wan, X., Dai, C., Du, X., Chen, X., & Guizani, M. (2018). Security in mobile edge caching with reinforcement learning.

## ADDITIONAL READING

Abomhara, M. & Køien, G. M. (2014, May). Security and privacy in the Internet of Things: Current status and open issues. In *2014 International Conference on Privacy and Security in Mobile Systems (PRISMS)* (pp. 1-8). IEEE.

Alsheikh, M. A., Lin, S., Niyato, D., & Tan, H. P. (2014). Machine learning in wireless sensor networks: Algorithms, strategies, and applications. *IEEE Communications Surveys and Tutorials*, *16*(4), 1996–2018. doi:10.1109/COMST.2014.2320099

Alsheikh, M. A., Lin, S., Niyato, D., & Tan, H. P. (2018). Machine learning in wireless sensor networks: Algorithms, strategies, and applications. *IEEE Communications Surveys and Tutorials*, *16*(4), 1996–2018. doi:10.1109/COMST.2014.2320099

Alsumayt, A. John Haggerty., & Ahmad Lot. (2016). Detect DoS attack using MrDR method in merging two MANETs. IEEE.

Arsalan Mohsen Nia., & Niraj K. Jha. (2016). A Comprehensive Study of Security of Internet-of-Things. *IEEE Transactions on Emerging Topics in Computing*, 99.

Bharadwaj, A. Subramanyam, G., Dr., Vinay Aasthi, Dr., & Hanumat Sastry, Dr., (2016). Solutions for DDos attacks on cloud. IEEE.

Butun, I., Morgera, S. D., & Sankar, R. (2014). A survey of intrusion detection systems in wireless sensor networks. *IEEE Communications Surveys and Tutorials*, *16*(1), 266–282. doi:10.1109/SURV.2013.050113.00191

Butun, I., Morgera, S. D., & Sankar, R. (2014). A survey of intrusion detection systems in wireless sensor networks. *IEEE Communications Surveys and Tutorials*, *16*(1), 266–282. doi:10.1109/SURV.2013.050113.00191

Cañedo, J., & Skjellum, A. (2016). Using machine learning to secure IoT systems. *Proceedings of 14th Annual Conference on Privacy, Security and Trust (PST)*. 10.1109/PST.2016.7906930

Chelli, K. (2015). Security issues in wireless sensor networks: attacks and countermeasures. Proceedings. of World Congress on Engineering.

Chen, F., Deng, P., Wan, J., Zhang, D., Vasilakos, A., & Rong, X. (2015). Data mining for the internet of things: Literature review and challenges. *International Journal of Distributed Sensor Networks*, *11*(8), 431047. doi:10.1155/2015/431047

Curran, J. (2015). IoT Development Poses Security, Privacy Problems. NY-US. Retrieved April 20, 2015, from http://search.proquest.com/docview/1675863933?accountid=36155

Diro, A., & Chilamkurti, N. (2017). *Distributed attack detection scheme using deep learning approach for Internet of Things*. Future Generations Computing Systems.

Diro, A. A., & Chilamkurti, N. (2017). Distributed attack detection scheme using deep learning approach for Internet of Things. *Future Generation Computer Systems*.

Gartner Press Release. Retrieved November, 2014, from http://www.gartner.com/newsroom/id/2905717

Gubbi, J., Buyya, R., & Marusic, S. (n.d.). [Security in Wireless Sensor Networks: Issues and Challenges.]. *Marimuthu Palaniswami*.

Husamuddin, Md. & Mohammed Qayyum. (2017). Internet of things: A study on security and privacy threats. International Conference on Anti-Cyber Crimes (ICACC). 10.1109/Anti-Cybercrime.2017.7905270

Internet of Things Global Standards Initiative. ITU. Retrieved June 26, 2015, from http://www.itu.int/en/ITU-T/gsi/iot/Pages/default.aspx

Kasinathan, P., Pastrone, C., Spirito., M. A., & Vinkovits, M. (2013). Denial-of-service detection in 6LoWPAN based Internet of Things. Piscataway, NJ: IEEE.

Kim, H.-J., Chang, H.-S., & Suh, J.-J. (2016). A study on device security in IoT convergence. Piscataway, NJ: IEEE.

Mukhopadhyay, D. (2015). *PUFs as Promising Tools for Security in Internet of Things*. IEEE.

Narayanan, A. (2014). Impact of Internet of Things on the Retail Industry. PC Quest, Cyber Media Ltd.; Retrieved May 20, 2014.

Raj, A. B., Ramesh, M. V., Kulkarni, R. V., & Hemalatha, T. (2012). Security enhancement in wireless sensor networks using machine learning. *International Conference on High Performance Computing and Communications and IEEE International Conference on Embedded Software and Systems*, 1264-1269. 10.1109/HPCC.2012.186

Riahi, C. Y., Natalizio, E., Chturou, Z., Bouab, A. (2013). A Systematic Approach of IoT Security. Retrieved 2013, from https://hal.inria.fr/hal-00868362/document

Shabana, K., Fida, N., Khan, F., Jan, S., & Rehman, M. (2016). Security issues and attacks in Wireless Sensor Networks. *International Journal of Advanced Research in Computer Science and Electronics Engineering*, 5(7), 81–87.

Singh, V., Puthran, S., & Tiwari, A. (2017). Intrusion detection using data mining with correlation. *Proceedings of International Conference for Convergence in Technology*. 10.1109/I2CT.2017.8226204

Wind River Company. (2015). Security in Internet of Things. Retrieved January, 2015, from http://www.winddriver.com

Zeitouni, S., Oren, Y., & Wachsmann, C. (2016). Remanence decay SideChannel: The PUF Case. Piscataway, NJ: IEEE.

## KEY TERMS AND DEFINITIONS

**DoS:** Denial of Service.

**Flooding:** The deliberate congestion of communication channels through relay of unnecessary messages and high traffic.

**IoT:** Internet of Things.

**Jamming:** The communication channel between the nodes is compromised and occupied and jammed, thus preventing them the sensor nodes from communicating with each other.

**K-NN:** K-nearest Neighbor.

**Machine Learning:** A field of information technology that has the ability to learn data insights by using statistical techniques.

**Masquerade:** A malicious node may act as another legitimate node to capture the message in the network.

**RFID:** Radio Frequency Identification.

**SVM:** Support Vector Machine.

**WSN:** Wireless Sensor Networks.

## Chapter 4
# Effect of Channel Modeling on Intercept Behavior of a Wireless BAN With Optimal Sensor Scheduling

**Deepti Kakkar**
*Dr. B. R. Ambedkar National Institute of Technology, India*

**Gurjot Kaur**
*Dr. B. R. Ambedkar National Institute of Technology, India*

**Parveen Kakkar**
*DAV Institute of Engineering and Technology, Jalandhar, India*

**Urvashi Sangwan**
*Dr. B. R. Ambedkar National Institute of Technology, India*

## ABSTRACT

*Body area networks (BANs), a type of Personal Area Networks (PANs), form a significant part of health care applications. This chapter analyzes the effect of channel modeling on the intercept behavior of a wireless BAN while taking optimal sensor scheduling into account. A comparison is drawn between Lognormal and Weibull models for this case. Wireless BANs represent wireless networks of sensors allocated on, in, and around the human body. BANs are basically meant for health care applications where long-lasting and reliable operation is a must. Some healthcare applications carry sensitive information, therefore security is an important issue. A BAN with a sink node and various sensors is considered here along with an eavesdropper. Due to the radio wave propagation's broadcast nature, the wireless communication can be overheard by the eavesdropper. To safeguard the BAN, the propagation channels need to be characterized and modeled for designing reliable communication systems.*

DOI: 10.4018/978-1-7998-0373-7.ch004

## INTRODUCTION

This chapter begins with an illustration of Body Area Networks, their significance, examples of their applications, the various challenges faced by these wireless networks and detailed description of the issues dealt in this research work. Besides that, it gives the organization of the chapter.

## Body Area Networks

The usage of wireless links is a fascinating replacement for wired cables coupling various wearable devices. Body area networks (BAN) are a type of personal area networks (PAN). BANs form a significant part of health care applications due to increase in medical expenses. Moreover, the patients suffering from chronic diseases who require only restorative observation need not be admitted to a hospital as this can be easily done via intelligent monitoring of the patient's body using wearable devices. According to Moore's law, the development of small and handheld devices which could be used for communication around human bodies was certain. An example of BANs is pacemaker. Other applications are smart pills for monitoring glucose, deciding on drug delivery and systems for sensing eye pressure. In the same way, wearable computing is an interesting application such as physiological/medical monitoring of temperature, heart rate, and blood pressure, etc. A BAN comprises of a number of sensors that are implanted inside the body, or are simply allocated on and around it along with a hub. The hub or the sink node can be placed on the body itself or alternately near it. The placement of on-body hub is generally done near the torso. BANs are meant for long-term use. To elongate the life of a BAN, low power sensors with a short range are employed. Other methods taken into account for long term usage of a BAN are communication using relays, controlling transmitting power of a sensor and adapting the link between sensors accordingly.

## Operational Scenarios for BANs

The various operational scenarios for BANs can be:

- **On-Body:** As the name suggests, in case of on-body BAN, communication will take place from one place on body surface to some other place on the body surface. It is the most widely used operational scenario.
- **In-Body:** The transmissions are generally done from a sensor implanted inside the body to one which is placed on the body. This is because doctors prefer to implant as few sensors inside a body as possible for the obvious

reasons. So even when an implanted sensor needs to communicate with a similar sensor, the preferred path is often via some on-body sensors which act as relays. The in-body BAN mostly works in coordination with on-body BAN.

- **Off-Body**: It comprises of wireless communication from a sensor placed on the body to a closely placed device (or vice versa) whose distance of employment should not exceed 3m.
- **Body-to-Body:** They support communication from one subject's body to another subject's body.

## Frequency Bands in BANs

IEEE Standard 802.15.6 is used for short-extent, wireless transmissions in the proximity of a human body (though not restricted to humans). This standard makes use of ISM band and certain other bands in combination with frequency bands in agreement with relevant communication and medical regulatory authorities. It lets devices to work on quite low transmit power in order to reduce the Specific Absorption Rate (SAR) into the patient's body to minimal levels while increasing the battery life (Fields,1997). It enhances Quality of Service (QoS), for instance, by providing emergency messaging (Smith et al., 2013).

## Need For Channel Modeling

Besides long-term usage, there are other challenges related to BAN and one such issue is reliability. A BAN needs to be reliable and in order to ensure reliability of a BAN, it is vital to analyze and implement the best suited channel model for a particular case. It is because the fluctuation in a channel varies significantly with the body state, movements, position of antenna, and the intercommunication between the antenna and the body as well as the surroundings. Taking into account typical characteristics in terms of both first order and second order statistics of the wireless channel, it is of utmost importance to do appropriate channel modeling for BAN so that the system can act according to such features. In present study, Lognormal channel modeling and Weibull channel modeling are taken into consideration.

## Lognormal Channel

Lognormal channel model is a general model which is used in a wide range of environments. This modeling is also found to be appropriate to giving a detailed account of the dominating effect of shadowing in BANs. In this case the wireless links between the nodes of the body area network are modeled as the Lognormal fading channels. If $X$ is a Lognormal distributed random variable whose location factor is $m$ and scale factor is $\sigma$, then the corresponding Probability Density Function (PDF) of $X$ is given by

$$f_X(x) = \frac{e^{(-(\ln x - u))^2}}{x\sigma\sqrt{2\pi}} \tag{1}$$

## Weibull Channel

Weibull distribution depicts another inference of the Rayleigh distribution. Let's say X and Y are independent and identically distributed random Gaussian variables with zero-mean, then the envelope of $Z = \left(X^2 + Y^2\right)^{\frac{1}{2}}$ has Rayleigh Distribution. On the other hand, when the envelope is represented by $Z = \left(X^2 + Y^2\right)^{\frac{t}{k}}$, then the respective Probability Density Function (PDF) has Weibull distribution which is given by

$$f_Z(z) = \frac{kz^{k-1}}{2\sigma^2} e^{\frac{z^k}{2\sigma^2}} \tag{2}$$

Where $k$ and $\sigma$ are the shape factor and the scale factor of Weibull distributed random variable Z. In this case, the wireless links between the nodes of the body area network are modeled as Weibull fading channels.

Weibull distribution is closely related to other probability distributions; it incorporates between exponential distribution (k = 1) and Rayleigh distribution (k = 2). Before explaining the meaning of different values of shape parameter k, failure rate is a term that should be known. Failure rate is the frequency with which an engineered product or a single component could fail. The failure is expressed in unit of time. The shape parameter k can be explained by the values below:

- k < 1 shows that the failure rate decreases over time. If the defective items are replaced early after their failure or the items are of exceptional quality.

- $k = 1$ shows that the failure rate remains constant over time. This indicates that some randomly occurring external event may be causing the failure or the mortality.
- $k > 1$ simply puts that the failure rate increases with time. This suggests that the components might fail with time or the "aging" process is a crucial factor in the component's life.

## Security in BANs

Besides the various benefits of BANs which include cost effectiveness, exemption from long hospital stays and continuous monitoring of patient without interrupting their day to day activities, there are certain issues faced by BANs such as Quality of Service (QoS), standardization, security and privacy challenges and those related to power supply. However, among these security and privacy concerns for BANs are quite crucial and therefore require special attention. The transmitted and the collected data in BANs plays a major role in identifying the type of illness and other problems by examining the symptoms as well as in treatment of that illness. Thus, it is of utmost importance to make sure that these data remain secure failing which can lead to wrong treatment of the patient and in worst case his/her death.

However, it is difficult to provide a precise and adaptable security mechanism to check malicious communications with WBANs. The broadcast nature of the wireless communication makes the BAN data vulnerable to eavesdropping and modification. In addition to that, representative channel features in BANs, for e.g. quite low value of Signal-to-Noise-Ratio (SNR), fixed power resource of body sensors, finite memory capacity and limitations in terms of communication and data processing ability make the prospect of security attacks in Wireless BANs (WBANs) more probable as compared to traditional Wireless Sensor Networks (WSNs). Besides, in WBANs, both security and performance of the system have equal significance, therefore, it is difficult to integrate a high-level security technique in these resource-limited networks. Hitherto, even though there are already various preliminary applications of WBANs that take QoS into account besides being energy efficient, researches related to data security and privacy challenges are scarce and prevailing solutions are far from being developed fully. The major security and privacy requirements to ensure the safety of WBAN system includes Data Confidentiality, Data Integrity, Data Freshness, Availability of the network, Data Authentication, Secure Management, Dependability, Secure Localization, Accountability, Privacy rules and compliance requirement (Al-Janabi et al., 2017).

In (Al-Janabi et al., 2017), various security threats and possible security solutions in WBAN systems are discussed. WBAN communication architecture is formed by three different levels. Tier-1: Intra-WBAN communication: interaction of the sensor confined around body of patient & information to next level transmitted via Personal Server (PS), Tier-2: Inter-WBAN communication: connect PS and the user via Access Points and Tier-3: Beyond-WBAN communication: A Smartphone can act as bridge between Tier 2 & Tier 3. In (Behera, Panigrahy, & Turuk, 2018), a biometric based light weight authentication protocol which guarantees secure communication in WBANs is proposed. Authentication is achieved by establishing session key between users and application server. The proposed method has three phases— initialization phase, registration phase, and login and authentication phase. Initialization phase is performed by the System Administrator (SA) who initializes the application server by choosing a secret key whereas the registration phase is performed by the users. After that, the SA provides unique identity to user and finally after the login phase, authentication is performed between the user and the application server by exchanging session key.

A new authentication scheme for WBANs is proposed in (Wang, & Zhang, 2015), using bilinear pairing. It was shown that the proposed scheme is able to withstand stolen verifier table attack and can provide unlinkability. In (Wu et al., 2016), a novel anonymous authentication method is presented for WBANs, which provides proof of security of proposed method under a random oracle model. It points out that scheme presented by (Wang et al., 2015) cannot withstand impersonation attack. Proposed security scheme consists of setup phase, registration phase and authentication phase. This security scheme can resist against various kind of security of attacks and performance of this scheme is much better as compared to (Wang et al., 2015). The major challenges in the design of WBSN are presented in (Mavinkattimath, Khanai, & Torse, 2019), like accuracy: sensors should provide accurate and real time information, information awareness: sensors should change their behaviour according to the situation of user and surroundings such as temperature, comfort and efficient response: It should be easy to use and very comfortable, energy efficient. Mavinkattimath et al. (2019) also provide solution to above challenges by considering use of lossless compressor which can reduce transmission data and use of encryption to provide privacy to data.

## Eavesdropping

Eavesdropping attack is a severe security risk to a WSN or BAN since this security attack is a precondition for other attacks. Traditional BANs comprise of wireless sensor nodes provided with antennas in all directions which leads to omni-directional broadcast of radio signals making BANs vulnerable to eavesdropping. The wireless sensor nodes are required to transmit their data to a sink node or hub which can be either placed on the body near the torso (in case of on-body area network) or alternatively near the patient's body (in case of off-body area network). However, an eavesdropper tries to overhear the communication between a sensor node and the sink node which can further lead to interception. So, there is a need to avoid eavesdropping or interception from taking place.

## Optimal Sensor Scheduling

As mentioned above, that due to the radio wave propagation's broadcast nature, the wireless communication can be overheard and intercepted by an eavesdropper. Taking this into account, there is a need for using a sensor scheduling scheme which can be implemented to maximize the security of a BAN.

In order to minimize the cost of communication, certain algorithms eliminate or lessen a sensor node's redundant data and keep away from forwarding the useless data. As sensor nodes can examine the information they direct towards the sink node, they can measure mean or directivity of sensed data from other nodes. For example, in case of applications concerned with sensing and monitoring, usually the sensor nodes in the vicinity of a node recording an environmental characteristic normally list much the same values. This type of data redundancy because of the geographical correlation between sensor monitoring encourages methods for data aggregation in a wireless network. Aggregation lowers the load of data traffic in the network which assists in reducing the energy requirement of sensor nodes. The algorithms used for data aggregation are called sensor scheduling scheme.

Optimal sensor scheduling (Zou & Wang, 2016) is one such method using which the sensor node having maximum secrecy capacity is chosen to transmit its data to the sink node, where secrecy capacity refers to the difference in channel capacities of main link and wiretap link. Here, main link is the wireless link between a sensor node and the sink node of BAN whereas wiretap link is the link between that sensor node and the eavesdropper. Also, channel capacity is calculated using Shannon's formula (Shannon, 1949).

## Objective of the Study

The present work is carried out to analyze the effect of using different channel models on the security of a BAN with the following objectives:

1.    To compare the intercept probability of BAN in case of Lognormal channel with that of Weibull channel for optimal sensor scheduling.

2. To analyze the performance of Lognormal distribution and Weibull distribution with variation in their parameters.

Since the number of sensors is limited in a BAN, hence performance of both the channel models w.r.t. the number of sensors employed has also been taken into consideration using the same scheduling scheme.

## BACKGROUND

## Cryptographic Techniques

Wireless sensor networks present advantages in various practical implementations such as BANs but are exposed to the possibility of being harmed by several security threats, such as eavesdropping and that of hardware tampering. Therefore, in order to protect the privacy, accuracy, and integrity of WSNs or BANs, effective security systems are required. With a view to gain secure communications among nodes, a lot many ways of dealing with these attacks utilize symmetric encryption. A number of key management techniques have been put forward in the direction of generating symmetric keys. An unconventional key management technique known as random seed distribution with transitory master key (RSDTMK) (Gandino, 2014) is discussed here, which follows the arbitrary dispensation of secret data and a transitory master key utilized to create pairwise keys. This method overcomes the main disadvantages of the earlier strategies based on these techniques. Besides, it performs better than the state-of-the-art protocols in the way that it always provides a high security level.

RSDTMK is a key management technique for a WSN with addition of sensor nodes without knowledge of distribution. An introductory version of this scheme was illustrated in (Gandino, 2009). This new approach is a hybrid technique which can be regarded a combination of both the random key implementation as well as the transitory master key families. RSDTMK depicts two differences: in the place of keys, RSDTMK implements seeds, i.e., the values taken into account to start a pseudorandom number generator where the magnitude of the distributed rings is effectively enlarged by making use of a basic permutation function on the seeds. In particular, each sensor node gains a ring comprised of seeds arbitrarily chosen from a pool. The pairs of sensor nodes search for common seeds, then exhibit a random number and utilize it with a permutation function, with a view to transform the seed. The purpose of this procedure is to enhance the number of probable keys in the network as compared to the number of seeds in the pool. The new seed which results from this mechanism is employed as the data for a pseudorandom function, which is then put into effect with master key (MK) to create the keys which will be further utilized by the nodes. Lastly within the initialization phase itself, each node deletes MK and the seeds. Thus, contrasting RSDTMK with a model random distribution technique, while the protocols possess identical size of ring and of pool, then the sum total magnitude of probable keys which are used by RSDTMK is considerable and the drawbacks of compromised secret data are less. In model transitory master key techniques, in case MK is compromised, then the rest of the keys are also compromised. On the other hand, in RSDTMK, an opponent can create only those keys which depend on the compromised seeds. The assessment of RSDTMK and its comparison with the state-of-the-art techniques depict that RSDTMK reduces the disadvantages caused by compromised secret data.

RSDTMK succeeds other key management techniques with corresponding advantages and drawbacks. One of these approaches was Random Key Distribution (RKD). RKD is based on arbitrary dispensation of keys which are chosen from a pool by each sensor node. Any two sensor nodes can communicate only in case they have a common key. This scheme although limits the disadvantage caused by a compromised node, however it also results in decreasing the connectivity in the wireless network, because, in case two neighboring sensor nodes do not possess a shared key, they cannot interact with each other. The random key distribution schemes are applicable to mobile wireless networks with scope of adding nodes without knowledge of deployment. The principal random key pre-distribution scheme is EG, (Eschenauer, Laurent, & Gligor, 2002). In this scheme, a pool consisting of keys is generated before the deployment. After that, a ring comprising of keys which are arbitrarily chosen from the pool is assigned to each sensor node. Within the phase of network initialization, each node makes sure that the other neighboring sensor nodes share at least a key with it. The estimates of number of keys in the pool p and that in a ring r act as deciding factors in determining the protocol performance. When the value of p is equivalent to r, each sensor node is capable of establishing a shared key with almost all of its neighboring sensor nodes, however, if even a single node is compromised, then a significant part of the network becomes insecure whereas if the value of r is too small, a sensor node is not able to communicate with all of its neighboring sensor nodes. Further, if p has a high magnitude, the number of links that an eavesdropper can compromise with the knowledge of one key, is reduced. An up gradation of this technique is portrayed by the composite Random Key Pre-distribution (Chan et al., 2003), which is known as QC. In QC protocol, two sensor nodes can communicate only in case they have at least q keys in common, and they create a new key by executing a hash function on the combination of all of the common keys. The q-composite technique is more resilient against a small number of captured nodes, however it is more prone to get attacked by an eavesdropper who captured various nodes.

## Artificial Noise Generation Approach

Shannon (Zou et al., 2015) put forward the idea of security at physical layer of a network and (Wyner, 1975) carried on his work further, in which Wyner developed a framework which has its basis on realizable secrecy rates for a conventional wiretap channel standard which comprises of an eavesdropper in addition to a source and a destination. The wireless link from a sensor node or source to the sink node or destination is called as main link. On the other hand, the one from that sensor node to the eavesdropper node is represented as wiretap channel or wiretap link. The difference between the channel capacity of main link and the channel capacity of wiretap link results in secrecy capacity(Leung-Yan-Cheong S, Hellman M.,1978) of that particular sensor node. The event of intercept occurs in case the channel capacity of wiretap link exceeds that of main link i.e. when the secrecy capacity becomes negative. Intercept probability reduces with an increase in the secrecy capacity (Zhou, Xiangyun, & McKay, 2010), (Zou et al., 2013). However, the fading mechanism in wireless transmissions along with the presence of obstacles badly affects the secrecy capacity.

The disseminating behavior of the wireless communication makes the transmission over this medium prone to eavesdropping attack. Consider a fading wireless environment between the sensor node and the sink node with a passive eavesdropper in their proximity. In case of artificial noise generation scheme, it is assumed that the transmitting sensor node and its amplifying relays have comparatively more antennas as compared to the eavesdropper. Here, the sensor node ensures secrecy of its transmitted data by making use of some of the power it has at its disposal in order to produce 'artificial signal', which will only degrade the eavesdropper's channel. There are two possibilities here are, one where the sensor node has multiple transmitting antennas, and the second where amplifying relays fake the effect of using multiple antennas.

In order to overcome eavesdropping, considerable research attempts have been directed towards increasing the secrecy capacity of the wireless transmissions using the artificial noise generation approach (Zhou, Xiangyun, & McKay, 2010), (Goel,

Satashu, & Rohit Negi, 2008), (Goeckel et al., 2011). The security techniques taking into consideration the addition of artificial noise lets the authorized transmitters to create a typically formulated interfering signal (also represented as artificial noise) in such a way that only the wiretap link is degraded by it, however the expected receiver remains uninfluenced by the artificial noise. As a result of this, the channel capacity of the wiretap link is deteriorated without having any influence on the main link's channel capacity, which leads to an improvement in the secrecy capacity. In (Zhou, Xiangyun, & McKay, 2010), (Goel, Satashu, & Negi, 2008) a number of antennas are employed for creating the artificial noise and it is suggested that the total number of antennas at the authorized node should be more than that at the intended receiver node to make sure that the main link is not affected by the produced artificial noise.

## Conventional Relay Selection Scheme

In (Zou et al., 2013), extra sensor nodes were put at disposal in order to relay the communication between the sensor node and sink node, such nodes are called as relay nodes. When a large number of relay nodes are present, (Zou et al., 2013) researched on the choice of relay that should be made to improve the security of wireless communication, where the relay node gaining the maximum secrecy capacity against eavesdropper node is chosen as the optimal relay to carry the source-sink transmissions. On one hand, the relay selection improves the physical-layer security of the wireless network, it makes use of supplementary relay nodes and also requires complex synchronization among locally implemented relays, and therefore, the complexity of the system increases.

In the conventional relay selection in (Bletsas et al., 2006), (Zou et al., 2010), (Ikki, Salama, and Ahmed, 2010)just the channel state information (CSI) of source-relay link and relay-destination link are considered, however in the considered optimal relay selection in addition to the CSI of these two-hop relay links, the CSI of the wiretap link is also taken into account. Taking into consideration amplify-and-forward (AF) and decode-and-forward (DF) relaying protocols, the considered optimal relay selection schemes are represented by P-AFbORS and P-DFbORS, respectively. The conventional AF and DF based optimal relay selection (i.e., T-AFbORS and T-DFbORS) and multiple relay combining (i.e., T-AFbMRC and T-DFbMRC) are represented as standard schemes. The derivation of closed-form expressions of probability of intercept for the PAFbORS and P-DFbORS besides the T-AFbORS, TDFbORS, T-AFbMRC and T-DFbMRC schemes in Rayleigh fading channels, shows that for both AF and DF protocols, the probability of intercept of considered optimal relay selection is always less as compared to that of the traditional optimal relay selection and multiple relay combining schemes, which justifies the advantage of optimal relay selection. Lastly, the diversity order performance of optimal relay selection schemes shows that the considered optimal and traditional optimal relay selection schemes both gain the same diversity order M irrespective of the relaying protocol, where M represents the number of relays.

## Optimal Sensor Scheduling Using Nakagami Fading Model

### Fading

In the 1920s, experiments were done on the nature of mobile communication at Very High Frequencies (VHF), at a frequency close to 50 Hz. The results of these experiments showed that wireless medium is very hostile in nature especially when it comes to urban areas. This is because after performing these tests, it came to light that the quality of a wireless signal can vary from 'excellent' to 'no signal at all'. Besides, it was noticed that when a vehicle is moved over just a few meters, it leads to large variations in the received field strength.

Multipath reception is a feature of indoor or mobile wireless links. This kind of signal consists of a large number of wireless signals which were reflected on their path to the receiver in addition to the direct line-of-sight (LOS) signal. In crowded places like cities, the LOS is mostly hindered by obstacles such as buildings and the received signal is often the combination of differently delayed versions of the LOS signal. The received signal is basically a result of interference to the original signal and hence the quality of originally transmitted signal is degraded on its way to the transmitter. Besides, if the receiver is in motion, further variations in the original signal will be introduced with location and time due to the change in relative phases of the reflected wireless signals. This will result in what is called as fading. Fading is nothing but the amplitude and phase changes introduced in the originally generated transmitted signal.

### Nakagami Distribution

In addition to Rayleigh and Rician fading, superior fading models have been put forward for describing wireless fading. Certain empirical data are better matched by the Nakagami distribution in comparison to other models.

Nakagami fading takes place for multipath scattering when the delay-time spreads with distinguished sets of reflected signals is comparatively large. The phase of any individual reflected signal in a set is arbitrary; however, the delay times are nearly the same for all signals. Therefore, the envelope of every signal cumulated in the set has Rayleigh distribution. It is assumed that the average time delay varies considerably among sets (or clusters). In case the delay times also considerably exceed the bit time of a digital signal, then the different clusters will result in severe intersymbol interference, thus the multipath self-interference then becomes equivalent to the scenario of co-channel interference by multiple incoherent Rayleigh-fading signals.

The Nakagami distribution or the Nakagami-m distribution is a probability distribution which is related to the gamma distribution. It has got two parameters, one is a shape parameter m $m \geq \dfrac{1}{2}$ and there is another parameter known as controlling spread parameter $\Omega > 0$.

The PDF of Nakagami distribution is given by $f_X(x) = \dfrac{2m^m}{\Gamma(m)\Omega^m} x^{2m-1} e^{\frac{-mx^2}{\Omega}}$

$\forall x \geq 0$ 

(3)

## Optimal Sensor Scheduling in Industrial Environment

Zou et al., (2013) introduced multiuser scheduling technique for enhancing the security of a wireless network at physical layer without the requirement of any additional power resources and illustrated the security enhancement of wireless networks with regards to the secrecy capacity and intercept probability. After that, (Zou et al., 2016) examined the sensor scheduling for an industrial wireless network comprising of a sink node (destination)and a number of sensors with an eavesdropper in their vicinity, distinguishing from the multiuser scheduling for wireless networks as investigated. In industrial WSNs, the wireless link is particularly made complex by the presence of obstacles (such as machinery), friction of metal and vibrations of engine. This encourages the consideration of a complicated fading model (i.e., Nakagami channel model) for realization of the wireless link in industry, in the place of a comparatively simple fading model (i.e., Rayleigh channel model) which was taken into account. This sensor scheduling exhibited some merits over the traditional relay selection as well as the artificial noise generation techniques with regards to decreasing the complexity of implementing the system and cutting the power cost. Particularly, in (Zou, Yulong, Xianbin Wang & Weiming Shen, 2013) and (Zou et al., 2014), extra sensor nodes were made available in the wireless network and implemented in order to relay the communication between the transmitting sensor node and receiving sensor node, which are also known as relay nodes. When a number of relay nodes are present in a network, the researchers inquired into the selection of relay for improving the security of wireless network, where the relay node having the highest secrecy capacity against eavesdropping is selected as the most suited relay to assist the source-destination transmissions. Even though the criteria for selecting relay studied in these works enhances the wireless security at physical layer, it depends on extra relay nodes and also demands complex synchronization among the relays distributed in space, which results in increasing complexity of the wireless system. Besides that, the artificial noise generation techniques were introduced in (Zhou, Xiangyun, & McKay, 2010), (Goel, Satashu, & Negi, 2008), (Goeckel et al., 2011) in order to increase wireless network's security by producing a specially designed artificial noise signal to confuse the eavesdropper where the desired destination remains unaffected. However, this adds to the cost of required energy resources when artificial noise production is considered as compared to the case of sensor scheduling, where a sensor that possesses the highest secrecy against eavesdropping is selected for transmission of data without using any extra energy resources. As wireless sensor nodes are mostly provided with limited batteries for usage, the energy is among the most expensive resources in industrial WSNs. This makes the sensor scheduling approach a better option in comparison with the traditional artificial noise techniques from prospect of saving energy.

However, as discussed in the introduction, Nakagami channel gives a poor performance when applied to a BAN. So, in our research work, we have considered the use of Lognormal channel and Weibull channel and compared the corresponding results.

## BAN SECURITY ENHANCEMENT

As discussed earlier, optimal sensor scheduling scheme outplays the artificial noise generation approach as well as the cooperative relay technique. However, (Zou & Wang, 2016) proposed this sensor scheduling method for industrial networks. In the present research work, this technique has been implemented on BANs. Besides, the effect of channel modeling is also observed w.r.t. the aspect of network's security. Two of the channels which are considered fit for BANs are taken into consideration, namely Lognormal channel and Weibull channel. The complete process is implemented in the present work to enhance the security of a BAN via optimal sensor scheduling scheme with different channel models. The whole procedure is explained step by step in detail in order to be understood easily.

## System Model

Take up an on-body area network (on-body wireless BAN channels are most frequently used) comprising of N sensors, a sink node and an eavesdropper node as presented in Figure 1. All the nodes are assumed to have only one antenna, main link is represented by solid lines and wiretap link is represented by dashed lines. The eavesdropper can be a legitimate user or an illegitimate user who wants to tap the data information of other users. Denote N sensors. There are a number of guidelines, or better say technical requirements for body area networks from IEEE 802.15.6 standard. They depict the operation of BANs and their influence on key features for channel modeling. The guidelines are: BANs can comprise of 256 nodes at the maximum and 2 nodes at the minimum. The sensor nodes can be placed on the chest, wrists, ankles and hips whereas the sink node or the hub is preferably placed near the torso. The bit-rates of BANs can vary from 10 kb/s to 10 Mb/s. The maximum transmitter power that can be radiated is supposed to be 0 dBm (or 1 mW), and all

*Figure 1. BAN on a male subject, illustrating gateway (hub), sensors and in-body, on-body and off-body links, adapted from (Smith et al., 2014)*

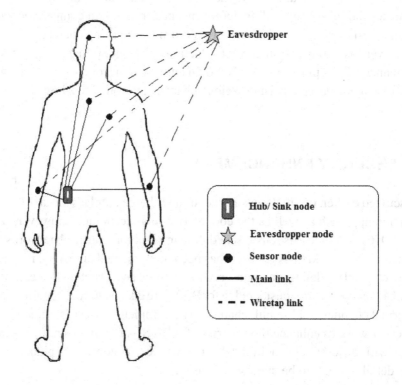

sensors should be capable of transmitting at 10 dBm (or 0.1 mW) i.e. the minimum transmitter power is 0.1 mW which by default fulfills SAR guideline of the Federal Communications Commission (FCC) of 1.6 W/kg in 1 g of body tissue. It should be feasible to add and remove nodes to/from the body area network in less than 3 seconds. There should be a scope for delay (at max 125 ms) and jitter (at max 50 ms) as per the IEEE standard. There should be a mechanism for saving power, for example, duty cycling. There should be a possibility of operation of at the very least ten co-located BANs which are randomly distributed in a volume of $6 \times 6 \times 6$ cubic metre. The standard distance between any two nodes is 2 m. Here, the frequency of transmission is 540 kHz at a carrier frequency of 2,360MHz with the sink node sampling frequency of 200 Hz.

Due to the dominating effect of shadowing in BAN, the use of lognormal channel is considered and it will be compared with Weibull channel in terms of the probability of intercept. The channel properties of both the wiretap link and the main link are taken into account in optimal sensor scheduling to effectively protect BAN against eavesdropping effect. In traditional scheduling method, only the channel properties of main link were considered and the sensor node with highest throughput was chosen for transmission to the sink node. In optimal sensor scheduling, it is assumed that the channel properties of both the wiretap link and the main link are known, it is a common supposition in the literature of physical layer security.

## OPTIMAL SENSOR SCHEDULING

Let us first calculate the secrecy capacity corresponding to each sensor. Consider that a sensor $s_i$ transmits its signal $x_i (E(|x_i|^2) = 1)$ with power $P_i$ and rate $R_i$ where $R_i$ refers to the maximum channel capacity from sensor $s_i$ to the sink.

Thus, the received signal at sink will be

$$y_s = \sqrt{P_i} h_{is} x_i + n_s \tag{4}$$

Where $h_{is}$ is the fading coefficient for the main link from $s_i$ sensor and $n_s$ refers to the AWGN noise with variance $N_0$. Applying the Shannon's channel capacity formula and using (4), the channel capacity of main link for $s_i$ sensor can be given as

$$C_s(i) = \log_2 \left( 1 + \frac{|h_{is}|^2 P_i}{N_0} \right) \tag{5}$$

Where $i \in S$. In the meanwhile, the eavesdropper overhears the signal transmitted from $s_i$ to the sink because of the broadcast feature of wireless communication. The eavesdropper endeavors to decode $x_i$. As already discussed, in the literature of physical layer security, the legitimate transmissions form sensor $s_i$ to sink, together with the modulation scheme and coding scheme, secret key and encryption algorithm (except the fact that signal $x_i$ is classified) is known to the eavesdropper. Hence, the signal which the eavesdropper overhears is given by

$$y_e = \sqrt{P_i} h_{ie} x_i + n_e \tag{6}$$

Where $h_{ie}$ is the fading coefficient for the wiretap link from $s_i$ sensor and $n_e$ refers to the AWGN noise with variance $N_0$. We can get the channel capacity of the wiretap link from sensor $s_i$ to eavesdropper $e$ in the same way as (5)

$$C_e(i) = \log_2 \left(1 + \frac{|h_{ie}|^2 P_i}{N_0}\right) \tag{7}$$

Therefore, using (5) and (7) the secrecy capacity of this wireless communication from sensor $s_i$ to sink, in the presence of an eavesdropper will be given as

$$C_{\sec recy}(i) = C_s(i) - C_e(i) \tag{8}$$

As the body area network consists of N sensors, so secrecy capacity is calculated for each sensor using (5), and the sensor having maximum secrecy capacity is selected for transmitting its data to the sink node. The optimal sensor is obtained by

$$OptimalUser = \arg \max_{i \in S} C_{\sec recy}(i)$$

$$= \arg \max_{i \in S} \frac{1 + \dfrac{|h_{is}|^2 P_i}{N_0}}{1 + \dfrac{|h_{ie}|^2 P_i}{N_0}} \tag{9}$$

Where $S$ is the set of N sensors. Classical channel estimation methods are used to acquire the channel state information, which is a requirement for deciding the optimal sensor. Firstly, each sensor calculates its own channel properties using channel estimation and then the estimated channel properties are transmitted to the sink. After analyzing all the sensors channel properties, the sink decides on the optimal sensor and notifies the whole BAN. Thus, the secrecy capacity of legitimate transmission based on the optimal sensor scheduling scheme can be expressed as

$$C_{\sec recy}^{optimal} = \max_{i \in S} \log_2 \left( \frac{1 + \frac{|h_{is}|^2 P_i}{N_0}}{1 + \frac{|h_{ie}|^2 P_i}{N_0}} \right) \tag{10}$$

## Exact Probability of Intercept Analysis

### 1. Lognormal Fading Channels

In this case the wireless links between the nodes of the body area network are modeled as the Lognormal fading channels. Thus, $|h_{is}|$ and $|h_{ie}|$ are lognormal distributed random variables whose respective location factors are m and k, and respective scale factors are $\sigma_{is}$ and $\sigma_{ie}$. Considering $X_{is} = |h_{is}|$ and $X_{ie} = |h_{ie}|$ the corresponding probability density functions PDFs of $X_{is}$ and $X_{ie}$ are given by

$$f_{X_{is}}(x_{is}) = \frac{1}{x_{is}\sigma_{is}\sqrt{2\pi}} e^{-\frac{(\ln x_{is} - m)^2}{2\sigma^2}}$$

$$f_{X_{ie}}(x_{ie}) = \frac{1}{x_{ie}\sigma_{ie}\sqrt{2\pi}} e^{-\frac{(\ln x_{ie} - k)}{2\sigma^2}} \tag{11}$$

The intercept probability of the transmission from sensor $s_i$ to the sink can be obtained as

$$P_{int}^i = \Pr\left[ C_{\sec recy}(i) < 0 \right] = \Pr\left[ C_s(i) < C_e(i) \right] \tag{12}$$

On substituting (5) and (7) into (12),

$$P_{int}^i = \Pr\left( |h_{is}|^2 < |h_{ie}|^2 \right) \tag{13}$$

As $|h_{is}|^2$ and $|h_{is}|^2$ are independent of each other, we get

$$P_{int}^i = \Pr\left(X_{is} < X_{ie}\right)$$

$$= \iint\limits_{x_{is} < x_{ie}} f_{X_{ie}}(x_{ie}) f_{X_{is}}(x_{is}) dx_{ie} dx_{is}$$

$$= \int_0^\infty \Gamma\left(m_i, \frac{m_i}{\sigma_{is}^2} x_{ie}\right) f_{X_{ie}}(x_{ie}) dx_{ie} \tag{14}$$

Where $\Gamma\left(m_i, \dfrac{m_i}{\sigma_{is}^2} x_{ie}\right)$ is the lower incomplete gamma function obtained as

$$\Gamma\left(m_i, \frac{m_i}{\sigma_{is}^2} x_{ie}\right) = \int_0^{\frac{m_i}{\sigma_{is}^2}} \frac{1}{\Gamma(m_i)} x^{m_i-1} \exp(-x) dx \tag{15}$$

From (10), we get the probability of intercept as

$$P_{int}^{optimal} = \Pr\left(C_{sec\,recy}^{optimal} < 0\right)$$

$$= \Pr\left[\max_{i \in S} \log_2\left(\frac{1 + \dfrac{|h_{is}|^2 P_i}{N_0}}{1 + \dfrac{|h_{ie}|^2 P_i}{N_0}}\right) < 0\right]$$

$$= \Pr\left[\max_{i \in S}\left(\frac{N_0 + |h_{is}|^2 P_i}{N_0 + |h_{ie}|^2 P_i}\right) < 1\right] \tag{16}$$

Where $S$ is the set of N sensors. As the random variables $|h_{is}|^2$ and $|h_{ie}|^2$ are independent of each other, (16) can be simplified as

$$P_{int}^{optimal} = \prod_{i=1}^N \Pr\left(\frac{N_0 + |h_{is}|^2 P_i}{N_0 + |h_{ie}|^2 P_i} < 1\right)$$

$$= \prod_{i=1}^N \Pr\left(|h_{is}|^2 < |h_{ie}|^2\right)$$

$$= \prod_{i=1}^{N} P_{int}^i \qquad (17)$$

Where $P_{int}^i$ is given by (14).

## 2. Weibull Fading Channels

Here the wireless links between the nodes of the body area network are modeled as Weibull fading channels. Thus, $|h_{is}|$ and $|h_{ie}|$ are Weibull distributed random variables whose respective shape factors are m and k, and respective scale factors are $\sigma_{is}$ and $\sigma_{ie}$. The corresponding equations to (14) for PDFs of Weibull channels are given by

$$f_{X_{is}}\left(x_{is}\right) = \frac{m}{\sigma_{is}}\left(\frac{x_{is}}{\sigma_{is}}\right)^{m-1} e^{-\left(x_{is}/\sigma_{is}\right)^m}$$

$$f_{X_{ie}}(x_{ie}) = \frac{k}{\sigma_{ie}}\left(\frac{x_{ie}}{\sigma_{ie}}\right)^{k-1} e^{-\left(x_{ie}/\sigma_{ie}\right)^k} \qquad (18)$$

Calculate the optimal intercept probability for Weibull model as per (17).

## RESULTS

The results are calculated by comparing Lognormal channel and Weibull channel on different types of variations when used in a BAN. The results obtained are in terms of probability of interception and number of sensors involved in constructing the BAN. Firstly, we will discuss the various parameters that are used for indication in graphs which will be followed by the graphs themselves, showing the results of the research work. Following is the elaborated detail of the work.

## Parameters

## 1. Intercept Probability

An eavesdropper or an unintended receiver tries to intercept the BAN signal from the network, thus, to calculate or measure how much close an eavesdropper is to interpreting the signal, we need a scale or a unit that can describe it. Therefore, for the required purpose we use the probability as the required quantity. The range of any probability is between 0 and 1. This scale simply describes that how likely the signal will be intercepted. A value of 0 means that signal cannot be intercepted and a value of 1 indicates that the signal is most likely to be intercepted. The parameter is used on the vertical axis of the graph against the other parameter which is on the horizontal axis.

2.    Number of Sensors

This parameter indicates the number of sensors considered in the BAN at the time of observation. The quantity of sensors will be varied from a single sensor to a maximum of 30 sensors at a time. This parameter is placed on the horizontal axis of the graph against the intercept probability that is taken on the vertical axis. This parameter will be a reference point on which the probability will be measured for a particular number of sensors.

3.    Shape Parameters

Shape parameters are used for random variable distribution of Weibull channel and Lognormal channel or any other channel. These parameters affect the shape of the distribution over the function variables. These parameters are unit less quantities with numeric value only and they describe the shape of a probability distribution which they are concerned with. It is not much related to the stretching or shifting of the probability distribution. Here, they are represented by m and k depending on the link considered. Further details are provided later in the text.

4.    Scale Parameters

Shape parameters are used for random variable distribution for Weibull channel and Lognormal channel or any other channel. These parameters are also portrayed by numeric values only. These are more concerned with the spread of the function. The larger the value of the parameter, the larger will be the spread of the probability distribution. Here, they are represented by $\sigma_{is}$ and $\sigma_{ie}$ depending on the link considered.

## Results

## Intercept Probability Comparison

The authors compare how the probability of interception in a BAN varies with the use of Lognormal channel model and Weibull channel model, when we take different numbers of sensors (N) into account. For this, we include a said number of sensors (say 5) and then calculate the intercept probability by using the probability equations discussed previously. Figure 2 shows the variation of intercepting probability with increasing number of sensors from 0 to 30. It clearly shows the comparison between the two channels. For a single sensor, the intercept probability for Weibull channel (graph a) is nearly 0.3 whereas that for Lognormal channel (graph b) is 0.1. When the number of sensors becomes 5, the corresponding intercept values are about 0.26 and 0.07, respectively. After this, the graphs gradually decline with an increase in the number of sensors involved in the construction BAN.

It should be noted that the graph is obtained by keeping values of m and k to 1.5 and $\sigma_{is} = \sigma_{ie} = 1$. Here, m and k represent the shape parameters of the main link (link between a sensor node and the sink node) and wiretap link (link between that sensor

*Figure 2. Comparison of Intercept Probabilities of Lognormal and Weibull channel*

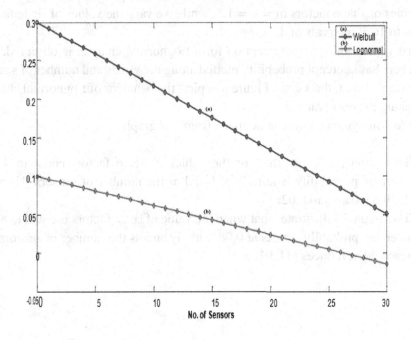

node and eavesdropper), respectively whereas $\sigma_{is}$ and $\sigma_{ie}$ are the corresponding values of scale parameters.

From this graph following observations can be made:

1.  Weibull channel is more susceptible to interception as compared to lognormal channel for a given number of sensors. This indicates that the lognormal channel should be preferred over Weibull channel as it is more resistive to interception and improves the overall resistance of the system to eavesdropping.
2.  Lognormal channel has zero probability of interception when the number of sensors (N) reaches the value of 30. This shows that, as the number of sensors involved in the procedure increase, the Lognormal channel can attain a state where chances of interception by anyone but the receiver are absolutely zero.

Hence, the result shows a significant difference in the performance of two channel models from the security point of view. Lognormal channel is more secure than the weibull channel because of its less vulnerability to interception.

## Individual Channel Behavior for Different Values of Scale Parameter

The values of shape factors m and k are kept the same (and equal to each other) and scale factors are varied for that particular value. The graphs are plotted by keeping the values of shape factors m = k = 1.5, while we vary the values of scale factors from 4 to 16 at intervals of 4.

Firstly, probability of interception for a Lognormal channel is observed. The graph here has intercept probability plotted along the y-axis and number of sensors (N) portrayed w.r.t. the x-axis. Figure 3 depicts the behavior of Lognormal channel for making the observations.

Following observations can be made from this graph:

*   The graph (a) shows that for the values of scale factors equal to 4, the intercept probability is initially 0.1 and as the number of sensors (N) reach 30, it decreases to 0.02.
*   The graph (b) illustrates that when the value of scale factors is equal to 8, the intercept probability starts at 0.08 initially and as the number of sensors (N) reach 30, it reduces to 0.018.

*Figure 3. Intercept probability for lognormal channel*

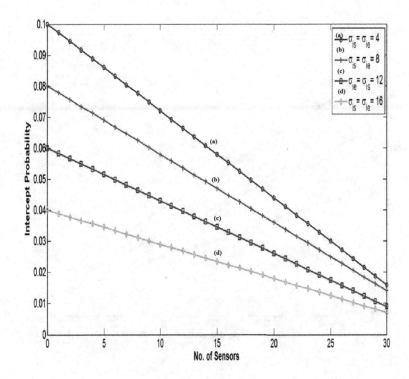

- The graph (c) depicts that for scale factors values equal to 12, the intercept probability starts at 0.06 initially and as the number of sensors (N) become 30 it falls to a value of 0.012.
- The graph (d) portrays that for scale factors values at 16, the intercept probability starts at 0.04 initially and as the number of sensors (N) increase to 30, it diminishes to 0.009.

Now, the probability of interception in the case of Weibull channel is determined. The graph here has intercept probability plotted along the y-axis and the number of sensors (N) is represented along the x-axis. Figure 4 depicts the behavior of Weibull channel for making the observations.

Following observations can be made from this graph:

- The graph (a) shows that for scale factors values equal to 4, the intercept probability starts at 0.3 initially and as the number of sensors (N) reach 30, it reduces to 0.07.

*Figure 4. Intercept probability for Weibull channel*

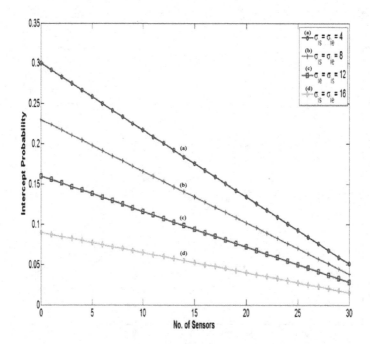

- The graph (b) shows that for scale factors values equal to 8, the intercept probability starts at 0.24 initially and as the number of sensors (N) reach 30 it reduces to 0.06.
- The graph (c) shows that for scale factors values equal to 12, the intercept probability starts at 0.16 initially and as the number of sensors (N) reach 30, it reduces to 0.05 respectively.
- The graph (d) shows that for scale factors values equal to 16, the intercept probability starts at 0.09 initially and as the number of sensors (N) reach 30 it reduces to 0.01 respectively.

This clearly depicts that increasing the values of scale parameters $\sigma_{is}$ and $\sigma_{ie}$ can help in reducing the overall chances of interception for any number of sensors whether 5 or 30 in case of both the channels. For direct comparison of the two channels and making the comparison easier, a table has been mentioned below that indicates the values for both the channels for N = 0 to N = 30, respectively. Table 1 depicts the comparison between the two channels.

From Table 1, following observations can be made:

- For σ=4, Lognormal channel has lesser probability of interception as compared to Weibull channel for a varying number of sensors.
- For σ=8, Lognormal channel has lesser probability of interception as compared to Weibull channel for a varying number of sensors.
- For σ=12, Lognormal channel has lesser probability of interception as compared to Weibull channel for a varying number of sensors.
- For σ=16, Lognormal channel has lesser probability of interception as compared to Weibull channel for a varying number of sensors

Thus, the detailed security analysis of the two channels for the specific case of BANs has been done in this work and is presented above. It can be said that overall, the Lognormal channel has better resistance to interception as compared to the Weibull channel when applied to a BAN and is more secure. Therefore, Lognormal channel should be preferred over Weibull channel for this particular application. Also, it is worth noticing that when the values of shape parameters are varied while keeping the values of scale parameters fixed, no change is observed in the intercept probability of the BAN in case of both the channels. So, this scenario has not been considered for plotting any graphs for the obvious reasons.

## LIMITATIONS

Besides the various advantages of the present work in terms of security enhancement in BAN, there are certain limitations to this research which are left to be dealt further. These are:

*Table 1. Comparison of both channels for different scale parameters*

| Channel | Lognormal Channel | | Weibull Channel | |
|---|---|---|---|---|
| Scale Parameter Value | N = 0 | N = 30 | N = 0 | N = 30 |
| 4 | 0.1 | 0.02 | 0.30 | 0.07 |
| 8 | 0.08 | 0.018 | 0.24 | 0.06 |
| 12 | 0.06 | 0.012 | 0.16 | 0.05 |
| 16 | 0.02 | 0.009 | 0.09 | 0.01 |

- It does not take into consideration the prioritization of sensors, for example, practically, certain sensor nodes may possess data which is time-critical i.e. with a stringent real time need, which must be given a higher priority as compared to the other sensor nodes in gaining access to the BAN channel. So, the Quality of Service (QoS) requirement is not considered in the present research work. Here all the sensor nodes have been assumed with equal priority and have been selected for transmitting data based only on their channel state information (CSI) without taking into account particular QoS needs for varied sensor data.
- Also, just a single-antenna scenario is examined; when each node of the network, whether it is a sensor node or the sink node is employed with a single antenna.
- In addition to the above limitations, due to the channel estimation errors, it is improbable to get the perfect value of CSI in order to do the sensor scheduling.

## FUTURE WORK

In order to overcome the above mentioned limitations, it is really important to devise ways in which the same levels of security can be attained without compromising the QoS requirements i.e. a sensor scheduling scheme need to be explored which guarantees each sensor's particular QoS need in addition to enhancing the BAN security.

Secondly, the results of this research work need to be generalized for a network having multiple antennas for each of its node. And lastly, the influence of CSI estimation errors on the intercept probability of a BAN needs to be investigated. All these challenging and interesting issues are left for future work.

## CONCLUSION

In this research work, the security performance of an on-body BAN is studied for two different channel models i.e. Lognormal channel and Weibull channel. The BAN is considered under stationary condition and the sensor scheduling approach used here is optimal sensor scheduling according to which the sensor having the maximum secrecy capacity is selected for transmission. The results demonstrate that:

- Lognormal channel outperforms Weibull channel in terms of BAN security i.e. the intercept probability for lognormal channel is less as compared to that of Weibull channel.
- The intercept probability for a BAN channel (whether Lognormal or Weibull) shows inverse proportionality with the value of scale factor or $\sigma_{ie}$ in case of both the channels i.e. the increase in the value of or $\sigma_{ie}$ will result in a decrease in the intercept probability.
- The intercept probability for a BAN channel decreases with an increase in the value of N i.e. number of sensors employed in the network in case of both the Lognormal channel as well as the Weibull channel.
- Lastly, the variation of shape factor m or k has no effect on the security parameters of a BAN i.e. the magnitude of intercept probability for either channel remains same irrespective of the value of m or k.

# REFERENCES

Al-Janabi, S., Al-Shourbaji, I., Shojafar, M., & Shamshirband, S. (2017). Survey of main challenges (security and privacy) in wireless body area networks for healthcare applications. *Egyptian Informatics Journal, 18*(2), 113–122. doi:10.1016/j.eij.2016.11.001

Behera, G., Panigrahy, S. K., & Turuk, A. K. (2018, February). A biometric based anonymous user authentication technique in wireless body area networks. In *2018 International Conference on Communication, Computing and Internet of Things (IC3IoT)* (pp. 308-312). IEEE. 10.1109/IC3IoT.2018.8668193

Bletsas, A., Khisti, A., Reed, D. P., & Lippman, A. (2005). A simple cooperative diversity method based on network path selection. *arXiv preprint cs/0510071.*

Chan, H., Perrig, A., & Song, D. (2003, May). Random key predistribution schemes for sensor networks. In IEEE Symposium on Security and Privacy, Oakland, CA, (pp. 197-213).

Cotton, S. L. & Scanlon, W. G. (2006, September). A statistical analysis of indoor multipath fading for a narrowband wireless body area network. In *2006 IEEE 17th International Symposium on Personal, Indoor, and Mobile Radio Communications.* (pp. 1-5).

Eschenauer, L., & Gligor, V. D. (2002, November). A key-management scheme for distributed sensor networks. In *Proceedings of the 9th ACM conference on Computer and communications security* (pp. 41-47). ACM. 10.1145/586110.586117

Fields, R. E. (1997). Evaluating compliance with FCC guidelines for human exposure to radiofrequency electromagnetic fields. *OET bulletin, 65*(10).

Fort, A., Desset, C., Wambacq, P., & Biesen, L. V. (2007). Indoor body-area channel model for narrowband communications. *IET Microwaves, Antennas & Propagation, 1*(6), 1197–1203. doi:10.1049/iet-map:20060215

Gandino, F., Montrucchio, B., & Rebaudengo, M. (2009, December). Random key pre-distribution with transitory master key for wireless sensor networks. In *Proceedings of the 5th International Student Workshop on Emerging Networking Experiments and Technologies* (pp. 27-28). ACM. 10.1145/1658997.1659012

Gandino, F., Montrucchio, B., & Rebaudengo, M. (2014). Key management for static wireless sensor networks with node adding. *IEEE Transactions on Industrial Informatics*, *10*(2), 1133–1143. doi:10.1109/TII.2013.2288063

Goeckel, D., Vasudevan, S., Towsley, D., Adams, S., Ding, Z., & Leung, K. (2011). Artificial noise generation from cooperative relays for everlasting secrecy in two-hop wireless networks. *IEEE Journal on Selected Areas in Communications*, *29*(10), 2067–2076. doi:10.1109/JSAC.2011.111216

Goel, S., & Negi, R. (2008). Guaranteeing secrecy using artificial noise. *IEEE Transactions on Wireless Communications*, *7*(6), 2180–2189. doi:10.1109/TWC.2008.060848

Leung-Yan-Cheong, S., & Hellman, M. (1978). The Gaussian wire-tap channel. *IEEE Transactions on Information Theory*, *24*(4), 451–456. doi:10.1109/TIT.1978.1055917

Mavinkattimath, S. G., Khanai, R., & Torse, D. A. (2019, April). A survey on secured wireless body sensor networks. In *2019 International Conference on Communication and Signal Processing (ICCSP)* (pp. 0872-0875). IEEE. 10.1109/ICCSP.2019.8698032

Shannon, C. E. (1949). Communication theory of secrecy systems. *Bell Labs Technical Journal*, *28*(4), 656–715. doi:10.1002/j.1538-7305.1949.tb00928.x

Smith, D., Hanlen, L., Miniutti, D., Zhang, J., Rodda, D., & Gilbert, B. (2008, October). Statistical characterization of the dynamic narrowband body area channel. In *First International Symposium on Applied Sciences on Biomedical and Communication Technologies. ISABEL'08.* (pp. 1-5). IEEE.

Smith, D. B., & Hanlen, L. W. (2015). Channel modeling for wireless body area networks. In *Ultra-Low-Power Short-Range Radios* (pp. 25–55). Cham, Switzerland: Springer. doi:10.1007/978-3-319-14714-7_2

Smith, D. B., Hanlen, L. W., Zhang, J. A., Miniutti, D., Rodda, D., & Gilbert, B. (2011). First-and second-order statistical characterizations of the dynamic body area propagation channel of various bandwidths. *Annals of Telecommunications*, *66*(3-4), pp. 187-203.

Smith, D. B., Miniutti, D., Lamahewa, T. A., & Hanlen, L. W. (2013). Propagation models for body-area networks: A survey and new outlook. *IEEE Antennas & Propagation Magazine*, *55*(5), 97–117. doi:10.1109/MAP.2013.6735479

Wang, C., & Zhang, Y. (2015). New authentication scheme for wireless body area networks using the bilinear pairing. *Journal of Medical Systems, 39*(11), 136. doi:10.100710916-015-0331-2 PMID:26324170

Wu, L., Zhang, Y., Li, L., & Shen, J. (2016). Efficient and anonymous authentication scheme for wireless body area networks. *Journal of Medical Systems, 40*(6), 134. doi:10.100710916-016-0491-8 PMID:27091755

Wyner, A. D. (1975). The wire-tap channel. *Bell Labs Technical Journal, 54*(8), 1355–1387. doi:10.1002/j.1538-7305.1975.tb02040.x

Zhen, B. (2008). Body area network (BAN) technical requirements. *15-08-0037-03-0006-ieee-802-15-6-technical-requirements-document-v-5-0. doc.*

Zhou, X., & McKay, M. R. (2010). Secure transmission with artificial noise over fading channels: Achievable rate and optimal power allocation. *IEEE Transactions on Vehicular Technology, 59*(8), 3831–3842. doi:10.1109/TVT.2010.2059057

Zou, Y., & Wang, G. (2016). Intercept behavior analysis of industrial wireless sensor networks in the presence of eavesdropping attack. *IEEE Transactions on Industrial Informatics, 12*(2), 780–787. doi:10.1109/TII.2015.2399691

Zou, Y., Wang, X., & Shen, W. (2013). Physical-layer security with multiuser scheduling in cognitive radio networks. *IEEE Transactions on Communications, 61*(12), 5103–5113. doi:10.1109/TCOMM.2013.111213.130235

Zou, Y., Wang, X., & Shen, W. (2013). Optimal relay selection for physical-layer security in cooperative wireless networks. *IEEE Journal on Selected Areas in Communications, 31*(10), 2099–2111. doi:10.1109/JSAC.2013.131011

Zou, Y., Wang, X., Shen, W., & Hanzo, L. (2014). Security versus reliability analysis of opportunistic relaying. *IEEE Transactions on Vehicular Technology, 63*(6), 2653–2661. doi:10.1109/TVT.2013.2292903

Zou, Y., Zhu, J., Wang, X., & Leung, V. C. (2015). Improving physical-layer security in wireless communications using diversity techniques. *IEEE Network, 29*(1), 42–48. doi:10.1109/MNET.2015.7018202

Zou, Y., Zhu, J., Zheng, B., & Yao, Y. D. (2010). An adaptive cooperation diversity scheme with best-relay selection in cognitive radio networks. *IEEE Transactions on Signal Processing, 58*(10), 5438–5445. doi:10.1109/TSP.2010.2053708

# Chapter 5
# AoSP–Based Secure Localization for Wireless Sensor Network

**Ankur Shrivastava**
*National Institute of Technology, Hamirpur, India*

**Nitin Gupta**
 https://orcid.org/0000-0001-5067-858X
*National Intitute of Technology, Hamirpur, India*

**Shreya Srivastav**
*National Institute of Technology, Hamirpur, India*

## ABSTRACT

*In wireless sensor network, node localization is helpful in reporting the event's origin, assisting querying of sensors, routing, and various cyber-physical system applications, where sensors are required to report geographically meaningful data for location-based applications. One of the accurate ways of localization is the use of anchor nodes which are generally equipped with global positioning system. However, in range-based approaches used in literature, like Angle of Arrival, the accuracy and precision decreases in case of multipath fading environment. Therefore, this chapter proposes an angle of signal propagation-based method where each node emits only two signals in a particular direction and knows its approximate position while receiving the second signal. Further, a method is proposed to define the coordinates of the nodes in reference to a local coordinate frame. The proposed method does the work with a smaller number of transmissions in the network even in the presence of malicious adversaries.*

DOI: 10.4018/978-1-7998-0373-7.ch005

## INTRODUCTION

Wireless Sensor Network (WSN) is a set of nodal sensor modules which connect to each other using wireless communication. WSNs can also be classified into: structured WSN and unstructured WSN (Maraiya et al., 2011). In the structured WSN, the sensor nodes are deployed in the network according to some pre-defined pattern giving an organized structure to the network. However, the unstructured WSN is just opposite to the structured. In unstructured WSN, the nodes are randomly deployed in the network. The unstructured WSN are mainly used in the region which is not accessible. In this undefined topology of the network, the sensor nodes have to maintain the topology dynamically by communicating with each other (Gupta et al., 2018). WSN is one of the significant area due to its major contribution in many applications. Main application areas for WSNs are healthcare applications, battlefield surveillance, forest fire detection, routing applications and monitoring environmental conditions (Fadel et al., 2015; Han et al., 2016; Rashid et al., 2016; Noel, 2017; Li, 2017). However, due to some of the limitations of the WSN, they are constrained in some areas as compared to traditional networks. These limitations of the WSNs are generally low power constraint, due to large number of applications it's heavy processing requirement and low memory constraint (Vijay et al., 2013).

Localization is one of the most important technology in WSNs (Chelouah, 2018; Boukerche et al., 2017; Halder & Ghosal, 2016; Chowdhury, 2016; Paul, 2017). Nodes in WSNs are randomly deployed, not planned & it is required to know their location to inform about events (Kumari, 2019; Tuna et al., 2017). This method of determining the location of the unknown nodes in the network is known as localization. Location of nodes provide various support for many location-based protocols such as routing algorithms to make efficient data routing decisions (Cadger et al., 2013; Ndiaye, 2017; Kumar, 2017). The location of the nodes can be found out using Global Positioning System (GPS) but it's not feasible for this case due to its large cost and large number of nodes (Xiao, 2016; Yassin, 2016). GPs unit is unable to work correctly inside buildings or dense forested areas and it may also expose to jamming and spoofing thereby creating a threat to many military applications (Bandiera et al., 2015). Basically, the localization is done using the information of the nodes whose location is already known or by using the communication information between the two unknown nodes in the network. The localization algorithm in WSNs has been classified in many ways. Zhu et al. (2014) has classified the localization as: direction-based approach and distance-based approach. In direction-based approach mainly the direction related parameters like received signal strength (RSS) are taken into account for the estimation (Azmi et al., 2018). Whereas, in direction-based approach directional parameters like Angle of Arrival (AoA) is considered. Angle of Arrival is defined to be the angle at which the signal has arrived. Mostly, the localization algorithms are mainly categorized into range based and range free. The range-based method employs range measurements, angle of arrival and angle of signal propagation. Angle of signal propagation is the particular angle at which the signal is send. Range free approach for localization is an economic approach; it does not provide high precision as much as range-based approach. It depends on the connectivity of two nodes which are fixed nodes. Some example of these range free methods is pattern matching based and hop-count-based localization techniques. Also, the range free approach does not need angle information or the range measurements and range free approach is favorable due its low cost and low energy utilization.

The rest of the chapter is organized as follows. Next section discusses the related work in the literature followed by the proposed scheme, performance metrics and conclusions.

## Related Work

Recently due to wide number of applications of localization in WSNs, it is a major attraction to be studied upon. Many studies and works have been done on this emerging technology.

Ghargan et al. (2016) have studied two approaches for finding the distance between the mobile node and the anchor node. These approaches aim for both outdoor and indoor environments. The first approach uses Log Normal Shadowing Model estimating parameters like path loss exponent and standard deviation. The second approach is the combination of Particle Swarm Optimization and Artificial Neural Network called as PSO-ANN algorithm. After comparison of both the approaches it is found that the distance error of mobile node is much improved in the PSO-ANN approach.

In the work done by Rong et al. (2006), they proposed Angle of Arrival based localization method. They derive it in noisy angle measurements assumption. They also assumed that from neighboring nodes, the unknown sensor nodes can detect angle of incident signal. Kułakowski et al. (2010) also propose the Angle of Arrival localization technique-based on Antenna Array in WSNs. To decrease the location errors they introduced heuristic weighting function. They proposed that each anchor node in WSN be equipped with an array of four antennas. They should be arranged in a square. Also, the algorithm is in decentralized structure, so there is no transmission from the anchor to the sensor nodes.

Tomic et al. (2016) discuss the localization problem in cooperative 3-D WSN. They proposed distribution algorithm for Received Signal Strength (RSS)/Angle of Arrival (AoA) localization problem. They use second order cone programming technique for non-convex nature of problem. It provides good localization accuracy in very little iterations. For the case of transmit power being different, the second order cone programming algorithm is more generalized. The proposed algorithm resulted better in terms of both the estimation accuracy and the convergence.

Pandey et al. (2016) proposed node localization for WSN with small world characteristics. This small world WSNs has reduced average path length (APL) between the nodes. For reducing APL long ranged links are introduced between the selected nodes in the WSNs and also it does not affect the clustering coefficients. For improvisation in localization accuracy this method comprises utilization of power of small world wireless sensor network characteristics and also uses the cognitive bandwidth. This localization accuracy obtained is better than the one obtained by conventional WSNs.

Ssu et al. (2005) describes a range free localization technique using mobile anchor points. Here each of the anchor nodes has its own GPS and the anchor node periodically broadcast its information in the sensing field. Each of the sensor nodes obtaining the location information from the anchor nodes can also compute their own location information and this computation is done locally. They also introduce certain enhancements like advance beacon point selection, chord selection and randomized beacon scheduling.

Vaghefi et al. (2015) propose a model for the cooperative sensor localization. This model is based on the asynchronous time of arrival measurements. For the estimation of the model, the problem domain is divided into two sub domains of localization and time synchronization. In this work maximum likelihood and Cramer-Rao lower bounds estimators are derived. Another method of semidefinite programming method is also estimated. A comparison work is also presented in the work between the estimators, which results that semidefinite programming method is better than the other methods in terms of the sensor localization accuracy.

Peng et al. (2015) proposed an approach including the optimization of the localization method and DV-Hop method based on the genetic algorithm for the location information of the unknown sensor nodes in the network. In this work a range free method of the localization is adopted by using the hop distance estimation. This work resulted that DV-Hop method when combined with the genetic algorithm concept provide better localization accuracy than the original DV-Hop algorithm.

Unlike various localization methods that have been proposed over the years for WSNs, the proposed Angle of Signal Propagation scheme makes use of Computational geometry. This scheme exploits location information of Anchor nodes to build a grid structure over entire sensor field, using hexagonal or square tiling. In the proposed scheme the random deployment of sensor nodes is considered for general nodes while grid construction for beacon. Query and location data forwarding is done through the shortest path between initial and first encountered anchor node. Alternatively changing the sector range of signals ensures that the query follow the shortest path between the two anchors. AoSP also does not have to wait for any other data thereby reducing any further delay. Moreover, this scheme is expected to be much more energy-efficient scheme compared to other methods used previously for localization in WSNs as it requires lesser number of transmissions where each node can send or receive only two signals.

## Proposed Work

### Introduction to the Problem

In this work, a novel technique is proposed for localization for a dense network. It is known that energy is the most crucial resource in WSNs, so it is tried to reduce communication as much as possible and on the contrary, a little computation is required. This is done in such a way that each general node is able to get its approximate location in exactly and only two communication signals (outgoing), i.e. a general node has to receive a maximum of two signals and also need to propagate two signals. Following assumptions are considered in this work:

1. The sensor nodes are static at their position, i.e. immobile in nature.
2. The node density is high, due to random deployment of nodes.
3. The assumption made here regarding the algorithm is that the sensor nodes are spaced equally between two anchor nodes.

*Figure 1. Hexagonal tiling*

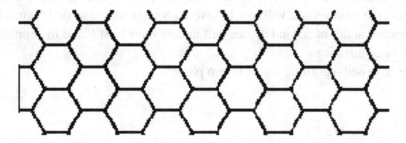

*Figure 2. Square tiling*

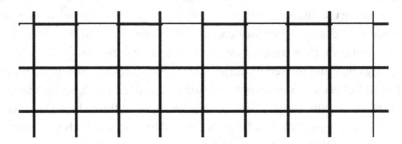

## Signal Propagation Based on Tiling

There are two types of regular tiling are considered with which a plane can be covered.

For the purpose of simulation, it has been assumed that the anchor nodes to be plotted according to some specific pattern, more exactly according to any one of the above-mentioned tiling e.g. if an anchor initiates phase-1, it will target its 24 or 16 neighbors in that many directions.

## Proposed Algorithm

The proposed algorithm works in two phases. After the first two phases, all nodes get their coordinates assigned. Most of them are assigned and validated twice. In the first phase the anchor nodes air the signal with a certain specified angle. An anchor node is the node which is assumed to know about its own location. The signal is aired in the specified range. This signal propagates along a certain direction in a zigzag manner. It counts the number of hops till it reaches another anchor node. Another anchor node assumes all the nodes on the path in a straight line. It calculates the path length between two anchor nodes & divides it by the number of hops. Hence, it evaluates the approximate coordinates of each of the node. It informs other nodes during its return trip. In the second phase, those general nodes, which have got two distinct locations due to the two different signals passing through them, will take average of these values, and will now, behave as anchor nodes of second degree. Now, these anchor nodes of second degree will initiate their First Phase by propagating signals in 24 directions.

The proposed algorithm works in two phases:

## First Phase

1.  An anchor node airs the signal with a certain angle & a specified range airing the signal in 24 directions or 16 directions according to the tiling.

    Here for Hexagonal Tiling as given in the Figure 3, the various ranges are defined as:

    ◦   0°-30° range
    ◦   15°-45° range
    ◦   ...
    ◦   345°-15°

    And for the square tiling, as given in the Figure 4 these ranges are:

    ◦   0° ± 15°
    ◦   22.5° ± 15°
    ◦   ...
    ◦   375.5° ± 15°

2.  The signal propagates along the direction specified in a zigzag manner as shown in the Figure 5 when stone is divided into in two parts using a wedge it gets a zigzag crack which approximates a straight line. Also, here the angle

*Figure 3. For the hexagonal tiling*

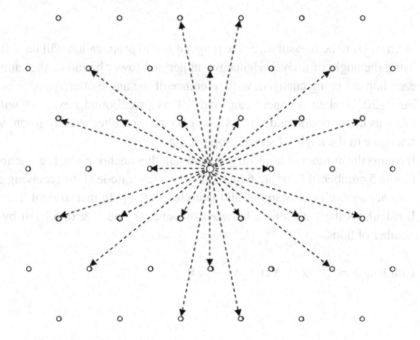

*Figure 4. For the square tiling*

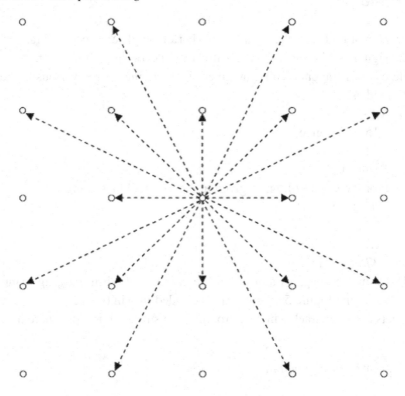

of arrival is to be measured, as the range of signal propagation will be defined using the angle of arrival. Hence the upper and lower bound of the range at each hop are being changed, with reference of the angle of arrival.

3.   First +20° is taken as upper bound and -10° as lower bound, next +10° will be taken as upper bound and -20° as lower bound, and alternatively so on, with reference to the angle of signal propagation.

4.   It counts the number of hops till it reaches another anchor node. E.g. for above Figure 5 number of hops are 6 including the anchor node at the receiving end.

5.   Another anchor node assumes all the nodes on the path in a straight line.

6.   It calculates the path length between two anchor nodes & divides it by the number of hops.

Path length $= \sqrt{((x1 - x2) - (y1 - y2))}$

One hop distance = Path length/ (no. intermediate nodes +1)

7.  Hence, it evaluates the approximate coordinates of each of the node, using computation geometry.
8.  Other nodes are informed during its return trip, which is the 2nd signal. It may also happen that some signals do not hit any anchor and pass through the whole field; such signals are considered as lost. Hence in total a general node has to receive 2 signals, and also propagate 2 signals only.

*Figure 5. Crack travelling in a brick wall*

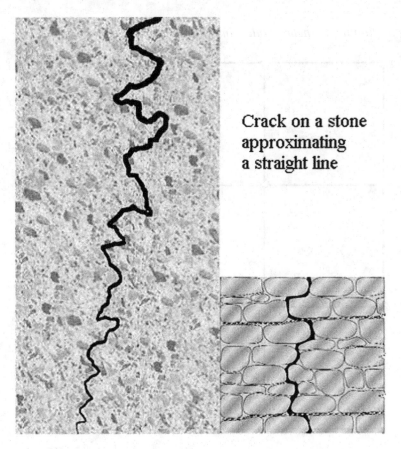

Crack on a stone approximating a straight line

## Second Phase

1.  Here the anchor nodes are plotted in a grid i.e. a square tiling, and they initiate the signal propagation as shown in the Figure 6 of signal propagation.
2.  Now, the signal may travel in a zigzag manner until they reach an anchor, and returning back on the same path, allotting all the nodes on the path an approximate value of its location. This completes Phase 1, for the original anchor nodes. The points shown with double circles are the original anchor nodes.
3.  Now, during Phase 1, some of the nodes have got their position twice because of intersection of two paths. Now, these general nodes will transform into anchor nodes of 2nd degree. And they will assume their position as the average of the two location values they have got. Original anchor nodes are shown with one circle and secondary anchor nodes are shown with double circles in Figure 8.

*Figure 6. Plotting of anchor nodes, initiation of signals by them*

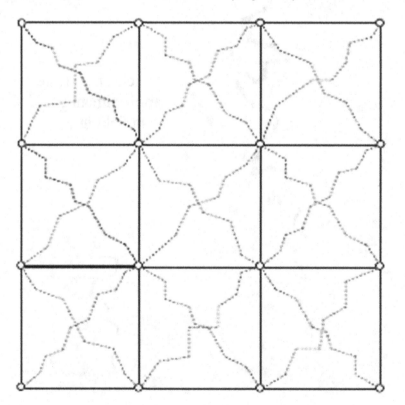

*Figure 7. Return of signal & assigning location*

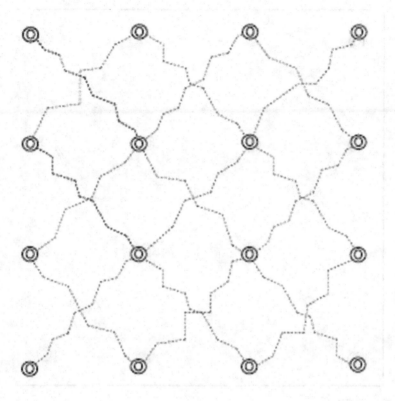

In Figure 9, double circles are Anchor node of 1st degree (original Anchors), triple circles are anchor node of 2nd degree and the fire sign indicates an anchor node of 3rd degree

4.  Now, here the anchor nodes are created up to the 3rd iteration. This iteration will go on up to have wholly covered the entire field. Here, it is shown that all signals strike the anchors but however as discussed, some signal do not hit anchors and pass throughout the field without striking. So many iterations have to be made and the last iteration will be when hop count is 1 or 2 as indicated in Figure 10.

For reducing the number of transmissions in this algorithm there following points may be considered:

1.  Only two signals are to be transmitted from a node.
2.  Only two signals are sensed per node in the network.
3.  Each node in the network makes only one computation.

*Figure 8. Assigning 2nd degree anchor nodes*

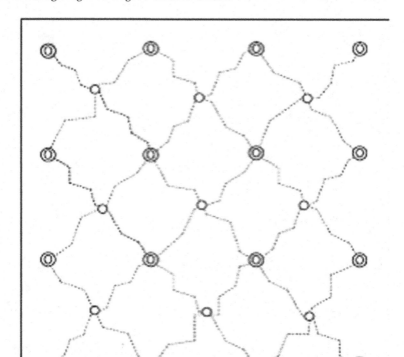

For improving the algorithm, the followings are included:

- Query and data forwarding is done through the shortest path and the shortest path is a straight line.
- To make the signal move in straight line: the sector range is alternatively changed from $[\theta + 20, \theta - 10]$ to $[\theta + 10, \theta - 20]$, just like a crack on the stone.

Some of the major advantages of using the AoSP algorithm are:

- This method of localization requires a smaller number of computations, less memory and less bandwidth as compared to the previous methods of the localization.

*Figure 9. Phase 2: Assigning 2nd degree anchor nodes*

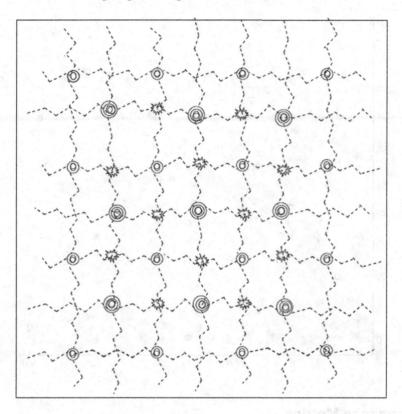

- As only 2 signal communication per node to be sent so there is a smaller number of transmissions in the network.
- Similarly, there are only two signal communication per node to be received again leading to lesser number of communications.

However, some of the limitations are:

- The performance of AoSP is low in sparse WSN as compared to the dense WSN.
- This algorithm does not estimate the exact coordinates of the nodes, it gives an approximate value.
- This method of localization requires directional antenna.

*Figure 10. Initiation by one of the anchor nodes in the sensor field*

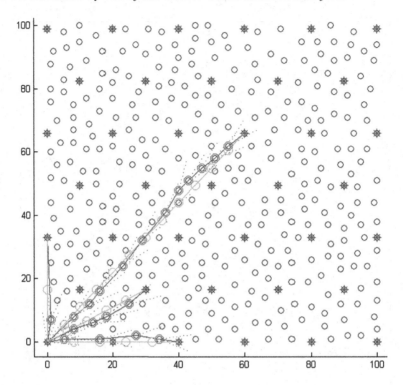

## Performance Metrics

- Total Energy Consumption: It is the average of the total energy consumption of all sensor nodes, which is always equal to 2 transmissions.
- Total Transmissions: Total transmissions are total number of packet transmissions during query and data forwarding is 2 packets always.
- Data Delivery Delay: Data Delivery Delay is the average time between transmission and reception of a packet, averaged over all paths is almost the time to travel between two nodes.
- Success Rate: Almost all of the nodes get their location and about 80% of the signals emitted complete their path at another anchor node.

## Security Mechanics

To protect the localization information, security mechanics explained in (Lazos, L., & Poovendran 2004) may be followed where before deployment, sensors and anchor nodes share a global symmetric key $K_0$. In future a study can be performed where

*Figure 11. Performance of our method*

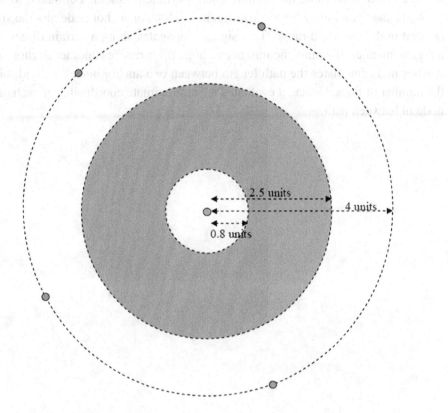

a wormhole link can be considered and effect of false hop count measurement on the proposed algorithm can be studied.

## Performance of the Method

- On an average the standard deviation ranges from 0.8 up to 2.5 units
- Minimum distance between any two nodes should be 4 units in a field of 100 x 100 units, i.e. it is determined that the node to be lying somewhere in the blue region.

## CONCLUSION

In this chapter, an angle of signal propagation-based localization scheme is discussed. This proposed scheme relies more on computation rather than communication. This scheme is expected to be much more energy-efficient scheme compared to other methods used previously for localization in WSNs. An anchor node airs the signal is aired in the specified range. This signal propagates along a certain direction in a zigzag manner. It counts the number of hops till it reaches another anchor node. Anchor node calculates the path length between two anchor nodes & divides it by the number of hops. Hence, it evaluates the approximate coordinates of each of the node in between path.

# REFERENCES

Azmi, N. A., Samsul, S., Yamada, Y., Yakub, M. F. M., Ismail, M. I. M., & Dziyauddin, R. A. (2018, July). A survey of localization using RSSI and TDoA techniques in wireless sensor network: System architecture. In IEEE *2nd International Conference on Telematics and Future Generation Networks (TAFGEN)* (pp. 131-136).

Bandiera, F., Coluccia, A., & Ricci, G. (2015). A cognitive algorithm for received signal strength-based localization. *IEEE Transactions on Signal Processing*, *63*(7), 1726–1736. doi:10.1109/TSP.2015.2398839

Boukerche, A., Oliveira, H. A., Nakamura, E. F., & Loureiro, A. A. (2007). Localization systems for wireless sensor networks. *IEEE Wireless Communications*, *14*(6), 6–12. doi:10.1109/MWC.2007.4407221

Cadger, F., Curran, K., Santos, J., & Moffett, S. (2013). A survey of geographical routing in wireless ad-hoc networks. *IEEE Communications Surveys and Tutorials*, *15*(2), 621–653. doi:10.1109/SURV.2012.062612.00109

Chelouah, L., Semchedine, F., & Bouallouche-Medjkoune, L. (2018). Localization protocols for mobile wireless sensor networks: A survey. *Computers & Electrical Engineering*, *71*, 733–751. doi:10.1016/j.compeleceng.2017.03.024

Chowdhury, T. J., Elkin, C., Devabhaktuni, V., Rawat, D. B., & Oluoch, J. (2016). Advances on localization techniques for wireless sensor networks: A survey. *Computer Networks*, *110*, 284–305. doi:10.1016/j.comnet.2016.10.006

Fadel, E., Gungor, V. C., Nassef, L., Akkari, N., Malik, M. A., Almasri, S., & Akyildiz, I. F. (2015). A survey on wireless sensor networks for smart grid. *Computer Communications*, *71*, 22–33. doi:10.1016/j.comcom.2015.09.006

Gharghan, S. K., Nordin, R., Ismail, M., & Ali, J. A. (2016). Accurate wireless sensor localization technique based on hybrid PSO-ANN algorithm for indoor and outdoor track cycling. *IEEE Sensors Journal*, *16*(2), 529–541. doi:10.1109/JSEN.2015.2483745

Gupta, N. & Dayal, N. (2018, August). Optimal cache placement by identifying possible congestion points in wireless sensor networks. In *International Conference on Wireless Intelligent and Distributed Environment for Communication* (pp. 161-170). Cham, Switzerland: Springer. 10.1007/978-3-319-75626-4_12

Halder, S., & Ghosal, A. (2016). A survey on mobility-assisted localization techniques in wireless sensor networks. *Journal of Network and Computer Applications*, *60*, 82–94. doi:10.1016/j.jnca.2015.11.019

Han, G., Jiang, J., Zhang, C., Duong, T. Q., Guizani, M., & Karagiannidis, G. K. (2016). A survey on mobile anchor node assisted localization in wireless sensor networks. *IEEE Communications Surveys and Tutorials*, *18*(3), 2220–2243. doi:10.1109/COMST.2016.2544751

Kułakowski, P., Vales-Alonso, J., Egea-López, E., Ludwin, W., & García-Haro, J. (2010). Angle-of-arrival localization based on antenna arrays for wireless sensor networks. *Computers & Electrical Engineering*, *36*(6), 1181–1186. doi:10.1016/j.compeleceng.2010.03.007

Kumar, A., Shwe, H. Y., Wong, K. J., & Chong, P. H. (2017). Location-based routing protocols for wireless sensor networks: A survey. *Wireless Sensor Network*, *9*(1), 25–72. doi:10.4236/wsn.2017.91003

Kumari, J., Kumar, P., & Singh, S. K. (2019). Localization in three-dimensional wireless sensor networks: A survey. *The Journal of Supercomputing*, 1–44.

Lazos, L., & Poovendran, R. (2004, October). SeRLoc: Secure range-independent localization for wireless sensor networks. In *Proceedings of the 3rd ACM workshop on Wireless security* (pp. 21-30). ACM. 10.1145/1023646.1023650

Li, X., Li, D., Wan, J., Vasilakos, A. V., Lai, C. F., & Wang, S. (2017). A review of industrial wireless networks in the context of Industry 4.0. *Wireless Networks*, *23*(1), 23–41. doi:10.100711276-015-1133-7

Maraiya, K., Kant, K., & Gupta, N. (2011). Efficient cluster head selection scheme for data aggregation in wireless sensor network. *International Journal of Computers and Applications*, *23*(9), 10–18. doi:10.5120/2981-3980

Ndiaye, M., Hancke, G., & Abu-Mahfouz, A. (2017). Software defined networking for improved wireless sensor network management: A survey. *Sensors (Basel)*, *17*(5), 1031. doi:10.339017051031 PMID:28471390

Noel, A. B., Abdaoui, A., Elfouly, T., Ahmed, M. H., Badawy, A., & Shehata, M. S. (2017). Structural health monitoring using wireless sensor networks: A comprehensive survey. *IEEE Communications Surveys and Tutorials*, *19*(3), 1403–1423. doi:10.1109/COMST.2017.2691551

Pandey, O. J., Kumar, A., & Hegde, R. M. (2016, March). Localization in wireless sensor networks with cognitive small world characteristics. In *2016 Twenty Second National Conference on Communication (NCC)* (pp. 1-6). IEEE. 10.1109/NCC.2016.7561180

Paul, A., & Sato, T. (2017). Localization in wireless sensor networks: A survey on algorithms, measurement techniques, applications and challenges. *Journal of Sensor and Actuator Networks*, *6*(4), 24. doi:10.3390/jsan6040024

Peng, B., & Li, L. (2015). An improved localization algorithm based on genetic algorithm in wireless sensor networks. *Cognitive Neurodynamics*, *9*(2), 249–256. doi:10.100711571-014-9324-y PMID:25852782

Rashid, B., & Rehmani, M. H. (2016). Applications of wireless sensor networks for urban areas: A survey. *Journal of Network and Computer Applications*, *60*, 192–219. doi:10.1016/j.jnca.2015.09.008

Rong, P., & Sichitiu, M. L. (2006, September). *Angle of arrival localization for wireless sensor networks. In 2006 3rd annual IEEE communications society on sensor and ad hoc communications and networks* (Vol. 1, pp. 374–382). Piscataway, NJ: IEEE.

Ssu, K. F., Ou, C. H., & Jiau, H. C. (2005). Localization with mobile anchor points in wireless sensor networks. *IEEE Transactions on Vehicular Technology*, *54*(3), 1187–1197. doi:10.1109/TVT.2005.844642

Tomic, S., Beko, M., & Dinis, R. (2016). Distributed RSS-AoA based localization with unknown transmit powers. *IEEE Wireless Communications Letters*, *5*(4), 392–395. doi:10.1109/LWC.2016.2567394

Tuna, G., & Gungor, V. C. (2017). A survey on deployment techniques, localization algorithms, and research challenges for underwater acoustic sensor networks. *International Journal of Communication Systems*, *30*(17), e3350. doi:10.1002/dac.3350

Vaghefi, R. M., & Buehrer, R. M. (2015). Cooperative joint synchronization and localization in wireless sensor networks. *IEEE Transactions on Signal Processing*, *63*(14), 3615–3627. doi:10.1109/TSP.2015.2430842

Vijay, U. & Gupta, N. (2013, January). Clustering in WSN based on minimum spanning tree using divide and conquer approach. In Proceedings of World Academy of Science, Engineering and Technology 79, p. 578. World Academy of Science, Engineering and Technology (WASET).

Xiao, J., Zhou, Z., Yi, Y., & Ni, L. M. (2016). A survey on wireless indoor localization from the device perspective. [CSUR]. *ACM Computing Surveys, 49*(2), 25. doi:10.1145/2933232

Yassin, A., Nasser, Y., Awad, M., Al-Dubai, A., Liu, R., Yuen, C., & Aboutanios, E. (2016). Recent advances in indoor localization: A survey on theoretical approaches and applications. *IEEE Communications Surveys and Tutorials, 19*(2), 1327–1346. doi:10.1109/COMST.2016.2632427

Zhu, G., & Hu, J. (2014). A distributed continuous-time algorithm for network localization using angle-of-arrival information. *Automatica, 50*(1), 53–63. doi:10.1016/j.automatica.2013.09.033

## KEY TERMS AND DEFINITIONS

**Anchor Nodes:** The sensing nodes which already know their location through some method like GPS.

**Angle of Arrival:** Angle of Arrival is the angle at which the signal has arrived.

**Angle of Signal Propagation:** Angle of signal propagation is the particular angle at which the signal is send.

**Localization:** To know the exact location coordinates of the sensing nodes.

**Range Based Localization:** The range based method employs range measurements, angle of arrival and angle of signal propagation.

**Range Free Localization:** Range free approach for depends on the connectivity of two nodes which are fixed nodes.

**Wireless Sensor Network:** A network consists of sensing nodes and one or more sinks connected through wireless connection. Sensing nodes sense some particular event and disseminate collected data through multihop communication through various intermediate nodes to the sink.

# Chapter 6
# Secure Routing Challenges for Opportunistic Internet of Things

**Nisha Kandhoul**
*Netaji Subhas University of Technology, India*

**Sanjay K. Dhurandher**
*Netaji Subhas University of Technology, India*

## ABSTRACT

*Internet of Things(IoT) is a technical revolution of the internet where users, computing systems, and daily objects having sensing abilities, collaborate to provide innovative services in several application domains. Opportunistic IoT(OppIoT) is an extension of the opportunistic networks that exploits the interactions between the human-based communities and the IoT devices to increase the network connectivity and reliability. In this context, the security and privacy requirements play a crucial role as the collected information is exposed to a wide unknown audience. An adaptable infrastructure is required to handle the intrinsic vulnerabilities of OppIoT devices, with limited resources and heterogeneous technologies. This chapter elaborates the security requirements, the possible threats, and the current work conducted in the field of security in OppIoT networks.*

DOI: 10.4018/978-1-7998-0373-7.ch006

## INTRODUCTION

Internet of Things (IoT) (Atzori et al., 2010) is a global network of connected objects that can be accessed via internet. IoT network provides a system for collecting, analyzing and processing the data generated by the sensor-based devices. IoT embodies a huge number of technologies and connects a variety of things or devices via unique addressing approach and standard protocols are used for communication. These devices are capable of interaction with one another and cooperate with their neighbors for achieving certain common goals. It is an innovation for the future of communications and computing. The connected objects in IoT vary from a person with a heart monitor to any device with built-in-sensors, i.e. devices having IP address and having the capability of data collection and interchange over a network without any manual intervention. The technology embedded in these objects enable them to interact with the external environment and affects their decision making. The devices can be managed and controlled remotely because of their inherent capability of connecting to the internet. IoT devices interact among themselves by transmitting and gathering information, sensing the environmental parameters like temperature, pressure etc., thereby transmitting the same to other devices in their communication range for further processing and other actions. The future era of internet will support interactions among humans, human based societies and smart objects held by them.

The network connections can be broadly categorized into two types based on its topology: infrastructure-based connection and infrastructure-less connections that is, ad-hoc or opportunistic connection. The infrastructure-based connections use pre- established infrastructure like base stations, routers, access points and manage the data a centralized way. In contrast, infrastructure-less connections do not use any infrastructure and make use of short-range radio techniques like Bluetooth, RFID, Wi-Fi etc for building decentralized networks.

Opportunistic Networks (Pelusi et al., 2006) is a class of Delay Tolerant Networks (Fall, 2003) that perform routing of messages and data sharing by exploiting the human characteristics like mobility patterns, similarities among humans, their daily routines and interests. Opportunistic IoT(OppIoT) extends the concepts of Opportunistic networks by merging human users and their smart devices. OppIoT explores the social side of IoT networks whereby the data is shared among communities, formed on the basis of movement and opportunistic contacts between humans and their personal devices like mobile phones, wearable devices, vehicles etc. Figure 1 shows a basic OppIoT network comprising of some mobile devices, laptops, sensors and human beings.

*Figure 1. Opportunistic IoT*

The excessive heterogeneity of devices and vast scale of OppIoT (B. Guo et al., 2013) systems magnifies the security threats. User's data are insecure as they are exposed to attacks from uncertified users or devices present across the network. In OppIoT based systems, securing the routing procedure is a challenging task as the packet travels across varying network topology and uniformity must be maintained while routing the packets between the source and the destination. Thus, dynamic prevention, detection and isolation for the attacks is required. In context of security, anonymity of users and data integrity and confidentiality, needs to be ensured. Also, for preventing unauthorized access to the system, authentication and authorization mechanisms need to be designed. Privacy requirements for OppIoT networks include both data protection and user's personal information security. Thus, security and privacy of the routing process is very important and needs significant research contributions.

This chapter elaborates the work done in the field of security of OppIoT. The authors present an in-depth comparison of various secure routing protocols in literature, listing pros and cons of each. The security attacks addressed and the methods used for solution are also mentioned. Later in the chapter, the authors elaborate the security and privacy concerns for OppIoT networks. As in today's age everyone is surrounded by sensing devices, all the data is at stake. So, ensuring the security of data is very important. The security goals and different types of attacks possible on the network are then elaborated. Attacks can be classified as active or passive. In active attacks, the attacker actively makes modification to either data or the network. Whereas the aim of the passive attacks is to gather information about the user, data or network topology. The later sections of the chapter explain how content and context-oriented security are different. Ensuring the security of data is known as content-oriented security. The data protection is enabled using authentication and encryption of messages. Context-oriented security focuses on ensuring the privacy of users by protecting information like node identity, location etc.

Finally, the authors present extensive comparison of the results obtained from the simulation of SHBPR and RSASec protocols, proposed in literature, using ONE simulator. It is observed that for all the chosen performance metrics in an environment where packet fabrication and black hole attacks are implemented, RSASec outperforms SHBPR as shown in subsequent sections.

## BACKGROUND

This section elaborates various security and privacy-based routing schemes that have been proposed in the literature for IoT, OppNet and OppIoT networks. All the protocols have been extensively compared and presented in Table 1. The positives and negatives of all the protocols have been listed. The protocols have also been compared based on the technical approach used for providing security and the attacks addressed.

With the help of the comparative analysis given above it can be easily concluded that the routing in OppIoT networks is prone to several types of attacks. In heterogeneous environments like the one in OppIoT networks, securing the routing procedure plays a crucial role in the smooth and safe operation of the network. Designing a solution that is universal applicable to all the possible routing attacks in OppIoT networks is an unsolvable problem. This is due to the fact that almost all the malicious attacks have an distinctive mode of operation and preempting future attacks in the

*Table 1. Comparative analysis of secure routing protocols*

| Secure Protocol Title | Pro's | Cons | Solving Method | Attack Handled |
|---|---|---|---|---|
| An altruism-based trust-dependent message forwarding protocol (Kumar et al., 2017) | • Threshold based adaptive routing using altruism value ensuring node participation<br>• Adaptability to all mobility patterns | • Malicious behavior of the nodes is not addressed | • Dynamic social graph for computing social and psychological attributes | • Denial of Service attack |
| Cryptography-Based Misbehavior Detection and Trust Control Mechanism (Dhurandher et al., 2017) | • Trust based social security<br>• Cryptography based message security<br>• Backward linkage of malicious detection through trust depreciation | • Dependency on infrastructure<br>• Excessive computation | • Energy constraints not considered<br>• Node profiling into normal, spy and judge nodes | • Black hole attack<br>• Worm hole attack<br>• Masquerade attack |
| A game theory based secure model against Black hole attacks (Chhabra et al., 2017) | • Use of Potential Threat (PT) messages to make the network aware of malicious nodes<br>• Timely message delivery | • Time consumed by a node to deliver a message is to be computed in advance | • Evolutionary Game Theory | • Black hole attack |
| Trust-Based Security Protocol Against Black hole Attacks (Gupta et al., 2013) | • Simplicity of malicious node detection and routing algorithm based on trust | • Social group values are static and arbitrarily chosen | • Trust based routing | • Black hole attack |
| A Privacy-Preserving and Secure Framework for Opportunistic Routing (L. Zhang et al., 2016) | • Routing metric encryption<br>• Group signature-based approach<br>• Anonymous authentication<br>• Effective key management<br>• External Adversaries are handled | • Dependency on certificate authority<br>• Malicious behavior of certified nodes is not considered | • Group Signature<br>• Homomorphic encryption<br>• Lightweight key agreement scheme | • Message forging<br>• Black hole attack<br>• Message replay attack<br>• Eavesdropping attack |
| A routing defense mechanism using evolutionary game theory (Guo et al., 2016) | • A proactive defense mechanism based on ACK information | • Storage overhead as two lists are maintained based on ACK and message information<br>• Sorting lists continuously | • Evolutionary Game Theory | • Denial of Service attack<br>• Black hole attack<br>• Worm hole attack |
| SecOMN: Improved Security Approach (Padhi et al., 2016) | • Offloading computationally intensive tasks to the nearest cloudlets<br>• Combination of symmetric and asymmetric key encryption | • Dependency on high computation power cloudlets<br>• Message drop attack can't be handled | • Cyber Foraging | • Cipher text only attack<br>• Known plain text attack<br>• Factorization attack |
| Trust Management for SOA-Based IoT (Chen et al., 2016) | • Trust feedback taken from nodes with identical social interests<br>• Adaptive filtering-based trust protocol parameter tuning<br>• Scalable to large IoT systems | • Maintenance of three lists friend list, location list and a CoI list<br>• Only trust related attacks are handled | • Adaptive trust-based protocol | • Bad-mouthing attack<br>• Opportunistic service attacks<br>• Self-promoting attack |
| Social Power for Privacy Protected Opportunistic Networks (Distl and Neuhaus, 2015) | • Detection of preestablished social ties with nominal impact on networking and computing resources | • Use of Bloom filters to save key and social link adds overhead<br>• Denial of service is possible<br>• New friendship links are not considered | • Social link analysis | • Eavesdropping attack<br>• Sybil attack |

*continued on the following page*

*Table 1. Continued*

| Secure Protocol Title | Pro's | Cons | Solving Method | Attack Handled |
|---|---|---|---|---|
| Secure Opportunistic Large Array for Internet of Things (Ansel et al., 2016) | • Energy efficiency<br>• Survivability during network partitions<br>• Robustness against mobility<br>• Cooperative transmission where node nearest to the interrogator transmits the data | • Security issues are not handled | • OLA transmission scheme | • None |
| FaceChange: Attaining Neighbor Node Anonymity (Chen and Shen, 2017) | • Neighbor node anonymity<br>• Real ID based encountering information collection upon disconnection<br>• Trust based fine grain control over information exchanged | • Dependency on Trust Authority for assigning trust<br>• Assumes nodes are cooperative<br>• Computationally intensive<br>• Eavesdropping attack is possible | • Pseudonym based routing scheme | • Bad mouthing attack<br>• Sybil attack<br>• Replay attack |
| Supernova and Hypernova Misbehavior Detection Scheme (Dhurandher et al., 2017) | • Use of B+ tree for managing the large amount of messages reduces complexity | • Sorting the count list for finding supernova nodes<br>• Computation overhead<br>• Messages generated by nodes are calculated for detecting hypernova and supernova nodes | • Probabilistic declaration of a node as supernova<br>• Analytical approach for declaring certain supernova as hypernova nodes | • Supernova attack<br>• Hypernova attack |
| History-Based Secure Routing Protocol to Detect Blackhole and Greyhole Attacks in Opportunistic Networks (Dhurandher et al., 2016) | • Efficient even in case of reduction in count of messages propagated in the network | • Use of four tables for storing trust and forwarding time values<br>• Considers that black or gray nodes do not alter table values | • Behavior based prediction | • Black hole attack<br>• Grey hole attack |
| An Energy-Aware Trust Derivation Scheme with Game Theoretic Approach in Wireless Sensor Networks for IoT Applications (Duan et al., 2014) | • Reduce the consumption of energy at the nodes<br>• Minimum Latency under the premise of security assurance | • The effect of malicious nodes on trust computation scheme is not considered | • Mixed strategy Nash Equilibrium of game theory | • Bad mouthing attack<br>• Denial of service attack<br>• Selfish attack |
| GMMR: A Gaussian Mixture Model Based Unsupervised Machine Learning Approach for Optimal Routing in Opportunistic IoT Networks (Vashishth et al., 2019) | • Combination of context-aware and context-free routing protocols | • Does not consider the effect of misbehaved nodes and focuses only on optimizing the routing procedure | • Machine learning | • None |
| Delay aware Secure Hashing for Opportunistic Message Forwarding in Internet of Things (Krishna and Lorenz, 2017) | • Parameters like trust, delay factor and time taken for transaction are taken into consideration | • Computation intensive | • Elliptic curve digital signature algorithm | • Message fabrication<br>• Bulk message handling<br>• Duplicate message request |
| A Privacy-Preserving Message Forwarding Framework for Opportunistic Cloud of Things (X. Wang et al., 2018) | • Use of double layered cloud server<br>• Secure mobility prediction | • Key encryption incurs overhead<br>• Communication overhead | • Attribute based cryptography | • Sybil attack<br>• Drop for profit<br>• Data tamper attack |
| An Asymmetric RSA based Security Approach for Opportunistic IoT (Kandhoul et al., 2019) | • Data security<br>• Asymmetric encryption | • User privacy is not handled | • RSA algorithm for encryption | • Eavesdropping attack<br>• Packet fabrication<br>• Cryptographic attacks |

network can prove to be very difficult task. However, designing a solution having the capability of effectively addressing a bunch of the above mentioned attacks can prove to be a novel achievement. The protocols should be designed keeping in mind the set standards of security and privacy for routing protocols. The designed procedure should reduce the security impact on the network thereby yielding a satisfactory network performance. The solutions need to be platform independent that are universally applicable to varying type of devices and are able to ensure: access control, confidentiality of data, privacy of users, trustworthiness among devices, compliance with defined security and privacy policies.

## MAIN FOCUS OF THE CHAPTER

## Security and Privacy Concerns in OppIoT

Some of the data that are collected and exchanged among the devices could be highly valuable thus requiring appropriate protection. As humans are present within the network of smart devices, the implementation of sensing technologies by IoT systems pose a threat to the privacy of the individuals (J. Lopez et al., 2017). With the increasing count of sensing devices surrounding us, the users data is collected without their active participation and their security is put at stake without them being aware of it. So, user security is of utmost importance in the present day world. The subsequent sections will elaborate the security goals, various types of attacks possible in the network and will differentiate between content and context-based security approaches.

### Security Goals

If an OppIoT network satisfies the following security goals (D. Airehrour et al., 2016), then it is said to be secure.

- **Availability:** Availability is the rendering of network services to every node in the network thereby ensuring the durability of all network services even in the case of presence of malicious nodes in the network.
- **Authenticity:** Authenticity is a mechanism through which the nodes are enforced to identify themselves and prove their identity in the network. This is required to shield the network from masquerading nodes trying to disrupt the network or accessing vital data.

- **Confidentiality:** Confidentiality protects the information from getting revealed to the harmful sources. Unauthorized access by the malicious nodes is prevented, thus disabling them from accessing crucial information related to the routing procedure or messages exchanged, from any sensible node or while such information is getting disseminated.
- **Integrity:** Integrity assures that the destination node has received correct data that not been modified during routing either through collision or via a deliberate attempt of tampering by some malicious node. Received data should be as originally sent.
- **Non-Repudiation:** Non-repudiation requires a sending node to accept the data sent by it while a receiving node accedes the receipt of the same. None of the parties involved can renounce the knowledge of sending or receiving the information.

## Security Attacks

The security attacks that affect the networks can be broadly categorized as shown in Figure 2.

*Figure 2. Categorization of Security Attacks*

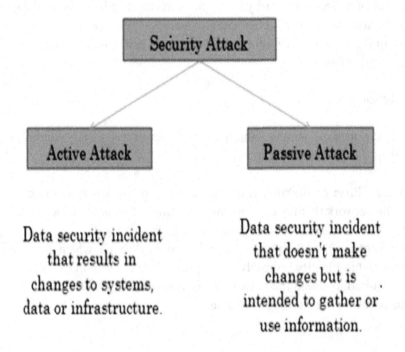

- **Passive Attacks:** These are data security incidents that do not create any changes in the network but are intended to gather or use information. Following are the instance of passive attacks (D. Kapetanovic et al., 2015):
- **Eavesdropping:** Eavesdropping is an attack whereby an unintended receiver reads the messages and conversations of others. Eavesdropping can act against the privacy protection of the network when the traffic conveys the control information about the network configuration.
- **Traffic Analysis:** Traffic analysis involves the interception of messages and scrutinizing them for deducing information from the communication patterns. Even encrypted messages provide high opportunity for analysis, thus causing harm to the network.


- **Active Attacks:** These data security incidents result in some changes to the system, data or infrastructure.
  Various types of active attacks include:
  - **Flooding attack:** It (P. Yi, Z. Dai, S. Zhang, Y. Zhong et al., 2005) is aimed at depleting the network resources like bandwidth, consuming the node's computational resources and battery power or breaching the routing operations to downgrade the performance of the network.
  - **Black-hole attack:** In this, a routing protocol is targeted by the attacker for advertising itself as possessing the shortest route to the destination for intercepting its packets. On receiving destination route request, the attacker creates a reply message for advertising itself as having a valid route to the destination. The attacker then drops or consumes (M. Salehi et al., 2015) the received messages without forwarding them.
  - **Grey-hole attack:** This is a selective packet drop attack where the attacker drops the messages with a certain probability. The attacker node executes such a malicious behavior for some time, then switches to its normal mode.
  - **Wormhole attack:** This is a form of collaborative attack (C. Tang et al., 2008) where two nodes despite being apart geographically are connected to each other. One of the attackers receives packets at one location in the network and then tunnels them to the other collaborating node thereby tampering the messages and disrupting the routing.
  - **Sybil attack:** A single malicious node presents multiple fake identities (S. T. Patel and N. H. Mistry, 2017) to participant nodes of the network. As the adversary appears in multiple places at the same time, it is able to attract maximum packets towards itself thereby manipulating them in numerous ways.

- ○ **HELLO flood attack:** "HELLO" packets (C. Adjih et al., 2005) are used by routing protocols for discovering neighboring nodes and establishing network topology. The attacker floods such hello messages in the network for exhausting node's resources thus protecting other messages from reaching the destination.

- ○ **Replay attack:** A malicious node records a benign node's valid control messages and sends them at some later time. This results in stale routes being recorded at other node's routing table. Replay attack can be used for disturbing the routing operations or for impersonating a specific node.

- ○ **Sinkhole attack:** The adversary node aims at attracting network traffic (C. Karlof and D. Wagner, 2003) to a particular compromised node by making it look promising to the surrounding nodes. The malicious nodes can then use all the gathered information for launching severe attacks in the network.

- ○ **Masquerade attack:** In masquerade attack, a node makes use of a fake network identity for gaining unauthorized access to the network. The malicious node then uses legitimate access identification for accessing other node's personal information.

- ○ **Denial of Service attack**: DoS attack (L. Liang et al., 2016) is used for influencing the network connection for making it non- accessible to its intentional users. The attacker floods the target node with traffic or sending unnecessary information to make it crash.

- ○ **Bad Mouthing attack:** This is a collaborative attack where the attackers collude for providing negative feedback about a selected victim node for lowering or destroying its network reputation (Z. Bankovi´c et al., 2011). This can considerably degrade the performance of the network.

- ○ **Free rider attack:** This is a selfish misbehavior where a node refuses to forward messages of other nodes but uses the resources of the network for forwarding its own messages.

- ○ **Supernova attack:** A flooding misbehavior is termed as supernova where a malicious node propagates random messages actively in the network. The supernova nodes consume valuable resources of the network like buffer, bandwidth and battery power by flowing malicious traffic in the network, thus making it unavailable for performing any useful work.

○ **Hypernova attack:** Hypernova nodes are those which generate and propagate random messages in the network and then deliver them to such nodes which are not existing in the network. This leads to the propagation of the messages in the network for even longer period of time than the one's initiated by the supernova nodes (S. K. Dhurandher et al., 2017).

## Content vs. Context Security

When it comes to the concept of security, one can discuss about data security (C. Castelluccia et al., 2009) that is content security or the privacy of users (context security), details of which are discussed below.

• **Content Oriented Security:** The data that is broadcasted across the network may carry some confidential information regarding valuable assets, businesses and individuals. As most of the data is transmitted in a broadcast manner across OppNets, the security of this data is of utmost importance. Content-oriented privacy will be enabled by securing these data from eavesdroppers and other above mentioned attacks. The protection of the data is enabled via authentication and encryption of messages. Homomorphic encryption is gaining huge popularity these days as it provides capability to the intermediate nodes for performing basic operations over the encrypted data as if it were in its original form. Homomorphic encryption (C.-M. Chen et al., 2012) can be performed in two ways:

○ **Symmetric-Key Homomorphism:** A single key is used for performing the encryption and decryption of messages. This approach is more prone to attacks as the single shared key can be easily compromised.

○ **Public-Key Homomorphism:** This approach uses two different keys at the sender and the receiver. This approach is more secure as compared to the one mentioned above.

Another approach to security is based on the use of trust. Trust (N. Karthik and V. S. Dhulipala, 2011) is the confidence level of one node on another node. While performing routing, a node is chosen as the next hop if its trust value is higher than others. But the major challenge faced in using encryption and trust based approaches is the dispersion of keys and trust to other nodes in the presence of malicious nodes in the network. In ordinary networks, the secrets are distributed during the initialization phase which are used later on.

But using this approach for highly dynamic scenarios with large number of security domains, like the ones in OppIoT, is infeasible. Thus, switching to public-key cryptography approach is mandatory. Public cryptography schemes use separate keys for performing encryption and decryption of messages and hence are more secure. Finally, considering active attacks is necessary. The adversary observes other users data and uses privacy mechanisms for modifying data without getting detected. Clearly, searching for a solution that balances the integrity and privacy is not an easy task. With advances in technology and a rise in the number of sensors around us, even the existence of messages in the network discloses a huge amount of information even after using secure encryption algorithms for securing their contents. Using the features of communications, an attacker can extract information from the (S. Pai et al., 2008) size and count of messages being transmitted, rate of message transmission, source and destination of transmissions etc. Such features cannot be concealed easily.

- **Context Oriented Privacy:** Context security (C. Boldrini et al., 2010) is the procedure of securing the personal information about the network nodes like location, domain name and so on. Context oriented privacy can be broadly categorized as:
  - **Identity Privacy:** The header of the packets carries information about the address of the source and destination involved in the communication process. So, the adversary can easily map the nodes to their geographical locations and thus easily linking the event messages to the region where they were created originally. Identifiers of the nodes must be changed regularly for preventing their identities from getting exposed to external observers and eavesdroppers. Using pseudonyms rather than their actual identities is one of the solutions for providing identity security of the nodes.
  - **Location Privacy:** The capability of keeping the location of nodes as secret in the network is referred as location privacy. The attacker in this case is a passive external node with a hearing range identical to the range of other nodes in the network. Such a type of attacker determines the angle of arrival of the messages and follows that direction for finding the sender of the messages. Once the location of the sender is disclosed, the eavesdropper can easily analyze the data packets sent and can further launch sophisticated attacks. Resolutions to this problem can be manifold. The messages can be sent using specific probability distribution function. Minimizing forged network traffic by malicious node isolation can be one of the solutions. Another way is scaling down the area to which the messages can be delivered.

Thus, it can be concluded that the need of the hour is to design routing protocols that provide a fine balance of content and context security. The user's personal information must be protected and the messages exchanged must be encrypted to provide data security to the users.

## EXPERIMENTAL RESULTS

The given section elaborates the comparison of the results obtained by the simulation of the protocols, SHBPR and RSASec, proposed in the literature. The simulation of the given approaches is conducted using the Opportunistic Network Environment(ONE) simulator (Keränen, A., Ott, J., & Kärkkäinen, T., 2009). The simulations are executed on real data trace of INFOCOM 2006, cambridge/haggle/imote traceset. This trace gives the contact pattern information between the associated nodes in network. The performance of the protocols is computed for metrics:-

- **Delivery Probability:** The probability of total count of messages getting successfully delivered to the receiver node within a said period of time.
- **Average Latency:** This is the measure of average delay incurred in the delivery of the message from the time of message generation to its final delivery to the destination.
- **Messages Dropped:** The count of total number of messages dropped at various nodes in the network.
- **Packet Delivery Percentage:** This metric is a composition of two parameters:
- **Modified Packets Received:** The ratio of altered packets received at the destination to the total packets received.
- **Correct Packets Received:** The ratio of unaltered or correct packets received to the total packets received at the destination.

The parameters chosen for executing the simulations are provided in Table 2. The Bluetooth granularity of scanning for a node is fixed at once per 120 seconds. The Time to Live(TTL) of the messages is 100 minutes and every simulation is executed for 337418 seconds. The message size is varied between 500 Kb to 1 Mb and the message creation occurs every 25-35 seconds.

The message fabrication attack is implemented in the network and the results are noted against varying degree of maliciousness in the network. The malicious nodes alter the message content by appending some random text to it. The count of malicious network nodes is varied from 5% to 25% and the result of this change is captured on the delivery of packets as illustrated in Figure 3. The number of

*Table 2. Simulation settings*

| Simulation Specification | Value |
|---|---|
| Area of simulation | 1000mt.*1000mt. |
| Range of Transmission | 10m |
| Power of Transmission | 15db |
| Size of Buffer | 100MB |
| Scan Response Energy | 0.08 Joules |
| Data set | Infocom2006 |
| Movement model | Stationary Movement |
| Simulation Period | 337418 seconds or 3.91 days |
| Type of Device | Imote |
| Granularity of Scanning | 120s |
| Transmission Energy | 0.5 Joules |
| Count of Participants | 98 |
| Count of Contacts | 170601 |
| Scan Energy | 0.06 Joules |
| Charging Coefficient | 20 Joules |
| Base Energy | 0.07 Joules |
| Threshold Energy | 5000 Joules |

modified packets received on an average for RSASec is 21.5% lower as compared to SHBPR. Also, the correct packets received are higher by 28.79% for RSASec.

Another attack implemented is the black hole attack and the results are noted for both the studied protocols for varying buffer size of the nodes and TTL of the messages.

Firstly, the time to live(TTL) of the messages is varied from 100 minutes to 300 minutes and the results for the performance metrics are given in Figure 4 to Figure 6. Figure 4 depicts the impact on the delivery probability as the TTL of the messages is varied. The lifespan of the messages rises as the TTL is increased, leading to an increase in the messages saved in the buffer. As the buffer is more occupied, there is an increase in messages dropped, thereby lowering the delivery probability of the messages. The average delivery probability of RSASec is 0.314% which is 7.53% higher as compared to SHBPR. The impact of varying TTL on the messages dropped is depicted in Figure 5. The average count of messages dropped

*Figure 3. Packet delivery percentage versus percentage of malicious nodes*

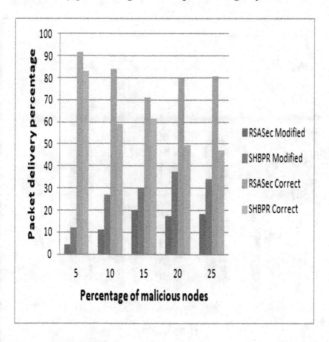

*Figure 4. Delivery probability versus TTL*

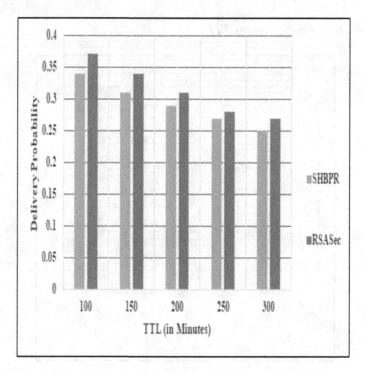

*Figure 5. Messages dropped versus TTL*

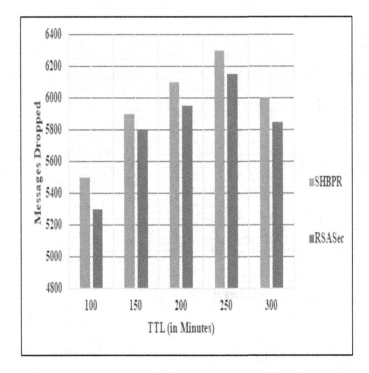

*Figure 6. Average latency versus TTL*

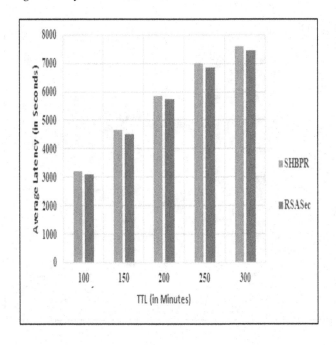

*Figure 7. Delivery probability versus buffer size*

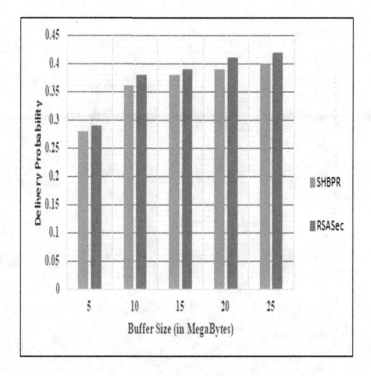

*Figure 8. Messages dropped versus buffer size*

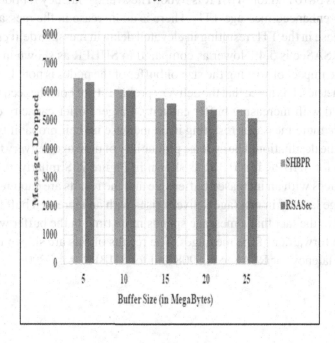

*Figure 9. Average latency versus buffer size*

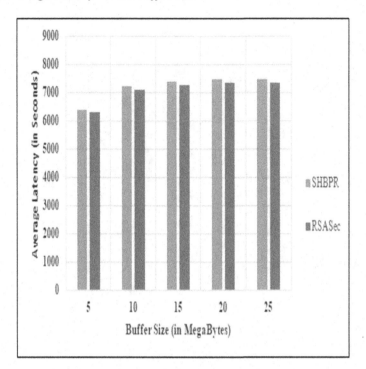

for RSASec is 5810 and for SHBPR is 5960. The average latency for both protocols rises with an enhanced message TTL. There is an increase in the message lifetime with an increase in the TTL resulting in elevated delay in message delivery. Average latency for RSASec is 5.4% lower as compared to SHBPR as shown in Figure 6.

Lastly, the impact of varying the size of buffer of the nodes is noted on the above mentioned metrics. It is noted that the delivery probability of both the secure protocols gets enhanced with increasing buffer capacity. Larger buffer capacity enables the nodes to store more messages resulting in an increase in their probability of getting delivered to the destination. The average probability of message delivery for RSASec is 0.378 and for SHBPR it is 0.362 as shown in Figure 7. Similarly, the messages dropped reduces with an increased buffer size and the results are captured in Figure 8. The average latency in message delivery rises with an increased buffer size. This is accounted to the fact that a message spends more time in the buffer waiting for a better secure forwarder of the messages. The results of this are shown in Figure 9. The average latency for RSASec is 6908 and for SHBPR it is 7290.

From the above results it can be concluded that RSASec outperforms SHBPR in terms of all the chosen performance metrics in an environment where packet fabrication and black hole attacks are implemented.

# CONCLUSION

With the increased connectivity to the internet and the technical revolution where everyone is surrounded by sensors, ensuring the security of the users and the messages exchanged by them is a major challenge for OppIoT networks. The user data is continuously being sensed and without their knowledge and active participation, the users privacy is at stake. So, the need of the hour is to design secure routing protocols which can be easily implemented across varying technologies and platforms like OppIoT that comprise of a wide spectrum of devices and human users. These security enhancement techniques must aim at mitigating as many attacks as possible, while achieving good performance in routing. The security techniques applied should not degrade the performance of the network and must consider the limited power and buffer availability in these devices. There is a huge scope for research in the field of security of OppIoT networks as not much work has yet been conducted in this field.

# REFERENCES

Adjih, C., Raffo, D., & Muhlethaler, P. (2005). Attacks against olsr: Distributed key management for security. *2ⁿᵈ OLSR Interop/Workshop,* Palaiseau, France, 14, pp. 1-5.

Airehrour, D., Gutierrez, J., & Ray, S. K. (2016). Secure routing for internet of things: A survey. *Journal of Network and Computer Applications, 66,* 198–213. doi:10.1016/j.jnca.2016.03.006

Ansel, V. P., Aboothahir, M. A., Smritilakshmi, A. S., & Jose, B. R. (2016, December). Secure opportunistic large array for Internet of Things. In *Proceedings 2016 Sixth International Symposium on Embedded Computing and System Design (ISED)* (pp. 201-204). IEEE. 10.1109/ISED.2016.7977082

Atzori, L., Iera, A., & Morabito, G. (2010). The internet of things: A survey. *Computer Networks, 54*(15), 2787–2805. doi:10.1016/j.comnet.2010.05.010

Bankovic, Z., Vallejo, J. C., Fraga, D., & Moya, J. M. (2011). *Detecting bad-mouthing attacks on reputation systems using self-organizing maps. Computational Intelligence in Security for Information Systems* (pp. 9–16). Springer.

Boldrini, C., Conti, M., Delmastro, F., & Passarella, A. (2010). Context-and social-aware middleware for opportunistic networks. *Journal of Network and Computer Applications, 33*(5), 525–541. doi:10.1016/j.jnca.2010.03.017

Castelluccia, C., Chan, A. C., Mykletun, E., & Tsudik, G. (2009). Efficient and provably secure aggregation of encrypted data in wireless sensor networks. [TOSN]. *ACM Transactions on Sensor Networks, 5*(3), 20. doi:10.1145/1525856.1525858

Chen, C. M., Lin, Y. H., Lin, Y. C., & Sun, H. M. (2012). RCDA: Recoverable concealed data aggregation for data integrity in wireless sensor networks. *IEEE Transactions on Parallel and Distributed Systems, 23*(4), 727–734. doi:10.1109/TPDS.2011.219

Chen, K., & Shen, H. (2017). FaceChange: Attaining neighbor node anonymity in mobile opportunistic social networks with fine-grained control. [TON]. *IEEE/ACM Transactions on Networking, 25*(2), 1176–1189. doi:10.1109/TNET.2016.2623521

Chen, R., Guo, J., & Bao, F. (2016). Trust management for SOA-based IoT and its application to service composition. *IEEE Transactions on Services Computing, 9*(3), 482–495. doi:10.1109/TSC.2014.2365797

Chhabra, A., Vashishth, V., & Sharma, D. K. (2017, March). A game theory based secure model against Black hole attacks in Opportunistic Networks. In *Proceedings 2017 51st Annual Conference on Information Sciences and Systems (CISS)* (pp. 1-6). IEEE. 10.1109/CISS.2017.7926114

Dhurandher, S. K., Kumar, A., & Obaidat, M. S. (2017). Cryptography-based misbehavior detection and trust control mechanism for opportunistic network systems. *IEEE Systems Journal*, (99), 1–12.

Dhurandher, S. K., Kumar, A., Woungang, I., & Obaidat, M. S. (2017). Supernova and hypernova misbehavior detection scheme for opportunistic networks. IEEE 31st International Conference on Advanced Information Networking and Applications (AINA). IEEE, pp. 387–391. 10.1109/AINA.2017.17

Dhurandher, S. K., Woungang, I., Arora, J., & Gupta, H. (2016). History-based secure routing protocol to detect blackhole and greyhole attacks in opportunistic networks. [Formerly Recent Patents on Telecommunication]. *Recent Advances in Communications and Networking Technology*, 5(2), 73–89.

Distl, B. & Neuhaus, S. (2015, January). Social power for privacy protected opportunistic networks. In *2015 7th International Conference on Communication Systems and Networks (COMSNETS)* (pp. 1-8). IEEE. 10.1109/COMSNETS.2015.7098697

Duan, J., Gao, D., Yang, D., Foh, C. H., & Chen, H. H. (2014). An energy-aware trust derivation scheme with game theoretic approach in wireless sensor networks for IoT applications. *IEEE Internet of Things Journal*, 1(1), 58–69. doi:10.1109/JIOT.2014.2314132

Fall, K. (2003). A delay-tolerant network architecture for challenged internets. *Proceedings of the 2003 Conference on Applications, Technologies, Architectures, and Protocols for Computer Communications*. ACM, pp. 27–34. 10.1145/863955.863960

Guo, B., Zhang, D., Wang, Z., Yu, Z., & Zhou, X. (2013). Opportunistic IoT: Exploring the harmonious interaction between human and the internet of things. *Journal of Network and Computer Applications*, 36(6), 1531–1539. doi:10.1016/j.jnca.2012.12.028

Guo, H., Wang, X., Cheng, H., & Huang, M. (2016). A routing defense mechanism using evolutionary game theory for delay tolerant networks. *Applied Soft Computing*, 38, 469–476. doi:10.1016/j.asoc.2015.10.019

Gupta, S., Dhurandher, S. K., Woungang, I., Kumar, A., & Obaidat, M. S. (2013, October). Trust-based security protocol against blackhole attacks in opportunistic networks. In *2013 IEEE 9th International Conference on Wireless and Mobile Computing, Networking and Communications (WiMob)* (pp. 724-729). IEEE. 10.1109/WiMOB.2013.6673436

Kandhoul, N. & Dhurandher, S. K. (2019). An asymmetric RSA based security approach for opportunistic IoT. *Proceedings 2ⁿᵈ International Conference on Wireless, Intelligent and Distributed Environment for Communication.* Springer, WIDECOM, Milan, Italy, pp. 47–60.

Kapetanovic, D., Zheng, G., & Rusek, F. (2015). Physical layer security for massive mimo: An overview on passive eavesdropping and active attacks. *arXiv preprint arXiv:1504.07154.*

Karlof, C., & Wagner, D. 2003). Secure routing in wireless sensor networks: Attacks and countermeasures. *Proceedings of the First IEEE International Workshop on Sensor Network Protocols and Applications.* IEEE, pp. 113–127. 10.1109/SNPA.2003.1203362

Karthik, N. & Dhulipala, V. S. (2011, April). Trust calculation in wireless sensor networks. In *2011 3rd International Conference on Electronics Computer Technology*, 4, pp. 376-380. IEEE. 10.1109/ICECTECH.2011.5941924

Keränen, A., Ott, J., & Kärkkäinen, T. (2009, March). The ONE simulator for DTN protocol evaluation. In *Proceedings of the 2nd International Conference on Simulation Tools and Techniques* (p. 55). ICST (Institute for Computer Sciences, Social-Informatics and Telecommunications Engineering). 10.4108/ICST.SIMUTOOLS2009.5674

Krishna, M. B. & Lorenz, P. (2017, December). Delay aware secure hashing for opportunistic message forwarding in Internet of Things. In Proceedings 2017 IEEE Globecom Workshops (GC Wkshps) (pp. 1-6). IEEE. doi:10.1109/GLOCOMW.2017.8269222

Kumar, A., Dhurandher, S. K., Woungang, I., Obaidat, M. S., Gupta, S., & Rodrigues, J. J. (2017). An altruism-based trust-dependent message forwarding protocol for opportunistic networks. *International Journal of Communication Systems, 30*(10), e3232. doi:10.1002/dac.3232

Liang, L., Zheng, K., Sheng, Q., & Huang, X. (2016). A denial of service attack method for an IoT system. In *Proceedings 8th International Conference on Information Technology in Medicine and Education (ITME)*. IEEE, pp. 360–364. 10.1109/ITME.2016.0087

Lopez, J., Rios, R., Bao, F., & Wang, G. (2017). Evolving privacy: From sensors to the Internet of Things. *Future Generation Computer Systems*, *75*, 46–57. doi:10.1016/j.future.2017.04.045

Padhi, S., Tiwary, M., Priyadarshini, R., Panigrahi, C. R., & Misra, R. (2016, March). SecOMN: Improved security approach for opportunistic mobile networks using cyber foraging. In *Proceedings 2016 3rd International Conference on Recent Advances in Information Technology (RAIT)* (pp. 415-421). IEEE.

Pai, S., Meingast, M., Roosta, T., Bermudez, S., Wicker, S. B., Mulligan, D. K., & Sastry, S. (2008). Transactional confidentiality in sensor networks. *IEEE Security and Privacy*, *6*(4), 28–35. doi:10.1109/MSP.2008.107

Patel, S. T., & Mistry, N. H. (2017). A review: Sybil attack detection techniques in WSN. In *Proceedings 4th International Conference on Electronics and Communication Systems (ICECS)*. IEEE, pp. 184–188. 10.1109/ECS.2017.8067865

Pelusi, L., Passarella, A., & Conti, M. (2006). Opportunistic networking: Data forwarding in disconnected mobile ad hoc networks. *IEEE Communications Magazine*, *44*(11), 134–141. doi:10.1109/MCOM.2006.248176

Salehi, M., Darehshoorzadeh, A., & Boukerche, A. (2015). On the effect of black-hole attack on opportunistic routing protocols. In *12th ACM Symposium on Performance Evaluation of Wireless Ad Hoc, Sensor, & Ubiquitous Networks*. ACM, pp. 93–100.

Tang, C., & Wu, D. O. (2008). An efficient mobile authentication scheme for wireless networks. *IEEE Transactions on Wireless Communications*, *7*(4), 1408–1416. doi:10.1109/TWC.2008.061080

Vashishth, V., Chhabra, A., & Sharma, D. K. (2019). GMMR: A Gaussian mixture model based unsupervised machine learning approach for optimal routing in opportunistic IoT networks. *Computer Communications*, *134*, 138–148. doi:10.1016/j.comcom.2018.12.001

Wang, X., Ning, Z., Zhou, M., Hu, X., Wang, L., Hu, B., ... Guo, Y. (2018). A privacy-preserving message forwarding framework for opportunistic cloud of things. *IEEE Internet of Things Journal*, *5*(6), 5281–5295. doi:10.1109/JIOT.2018.2864782

Yi, P., Dai, Z., Zhang, S., Zhong, Y., & ... . (2005). A new routing attack in mobile ad hoc networks. *International Journal of Information Technology, 11*(2), 83–94.

Zhang, L., Song, J., & Pan, J. (2016). A privacy-preserving and secure framework for opportunistic routing in DTNs. *IEEE Transactions on Vehicular Technology, 65*(9), 7684–7697. doi:10.1109/TVT.2015.2480761

# Chapter 7
# Middleware Approach to Enhance the Security and Privacy in the Internet of Things

**Vikash**
*Indian Institute of Information Technology, Allahabad, India*

**Lalita Mishra**
*Indian Institute of Information Technology, Allahabad, India*

**Shirshu Varma**
*Indian Institute of Information Technology, Allahabad, India*

## ABSTRACT

*Internet of things is one of the most rapidly growing research areas. Nowadays, IoT is applicable in various diverse areas because of its basic feature i.e., anything would be available to anyone at anytime. Further, IoT aims to provide service in a pervasive environment, although different problems crop up when the researchers move towards pervasiveness. Security and Privacy are the most intense problems in the field of IoT. There are various approaches available to handle these issues: Architectural security, Database security, Secure communication, and Middleware approaches. This chapter's authors concentrate on middleware approach from the security and privacy perceptive. Middleware can provide security by separating the end user from the actual complex system. Middleware also hides the actual complexity of the system from the user. So, the user will get the seamless services with no threats to security or privacy. This chapter provides a brief overview of secure middlewares and suggests the current research gaps as future directions.*

DOI: 10.4018/978-1-7998-0373-7.ch007

## INTRODUCTION

Internet of Things (IoT) is a combination of various technologies including sensors, actuators, embedded systems, cloud computing, next-generation of cheaper and smaller devices, objects, and things. Moreover, the researchers define IoT on the basis of its common characteristics, which involve the objects in the IoT scenario should be instrumented and interconnected to process anything intelligently and should be available to the end users anyhow, anywhere, anytime in anyway. Further, on the basis of IoT characteristics world can be categorize in four fundamental building blocks i.e., Radio Frequency IDentification (RFID), Machine to Machine communication (M2M), Wireless Sensor Networks (WSN), and Supervisory Control and Data Acquisition (SCADA), as shown in Figure 1 along with Internet. Further, these technologies are the building blocks of IoT. So, IoT automatically inherits the features and challenges existing in these technologies. Moreover, IoT evolved with various challenges as we move towards pervasiveness like Heterogeneity, Interoperability, Security, Privacy, Reliability etc., along with the preexisting issues.

IoT is the glue that tightens these four pillars through a common set of characteristics, networking methodology, and an abstract software layer middleware platform. The authors lend an abstracted view of these pillars along with level of applicability chronologically in IoT, which are as follows:

**WSN (Internet of Objects):** RFID is a low-cost, disposable contactless smartcard. RFID use radio waves for transmission of data from electronic chip attached to object to a dedicated system via a reader for tracking and identification the

*Figure 1. Building blocks of IoT*

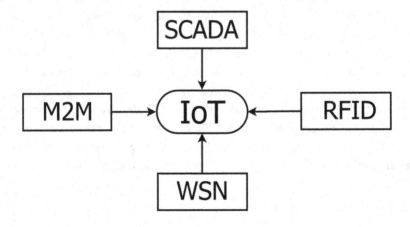

object. RFID tags are further categorized into two types of tags active and passive (Fujisaki, 2019).

**Active RFID Tags:** Active RFID tags are used to transmit the data continuously by using ID tags which are self-powered. These types of tags have their own battery, and use it to emit a signal with predefined strength in a time interval of about one second. Further, it can be divided into two types 1. Transponders awake and communicate when it come in the range of the reader. 2. Beacons used to broadcast its visibility continuously. So, the reader can communicate when it comes in the range.

**Passive RFID Tags:** These tags do not have any battery; the power is supplied by the reader for communication hence they have a long lifespan. Further, the radio waves from the reader are reflected to a passive RFID tag, the coiled antenna form a magnetic field within the tag, which draw a power from reader to energizing the circuits in the tag.

**M2M (Internet of Devices):** The Internet is just not only to communicate with people; it's more than the people percept. Nowadays, it is used intelligently to connect machines which must be able to communicate and interact with speeds, scales and capabilities beyond the original need and use of people. Further, IoT slowly unlock the various capabilities of Internet and making the world more agile and functional via M2M and other protocols. It uses devices (e.g., vehicle, mobile gadget) to collect the data through the networks and forward to central processing unit, which translate this data into meaningful information to provide knowledge about the system. Additionally, the technology has developed into the mobile environment to further improvement in various areas of communications (Althumali, 2018).

**WSN (Internet of Transducers):** WSN is mainly used to sense and collect data from sensors to monitor different environmental conditions, it includes various networks as VSN (visual or video sensor networks), vehicular sensor networks, BSN (body sensor networks), under water acoustic sensor networks (UW-ASN), interplanetary sensor networks (DTN), ubiquitous sensor network and others for information collection and processing (Keramatpour, 2017). Ubiquitous sensor network is a network of intelligent sensor network that could provide its services ubiquitously. WSNs need to be developed and deployed in the form of large number of sensors in various environments, remote area in ad-hoc manner. Further, middleware for WSN deals with the middle-level primitives between software and hardware to fill the gap.

**SCADA (Internet of Controllers):** SCADA is a type of industrial control system, which is used to monitor and control the industrial processes related to the physical world. Further, it is self-governing system based on a smart system concept or on close-loop control theory or a CPS that is used to connect, monitor and control devices via a network infrastructure (Katyara, 2017). This mainly concentrated on security aspect of IoT.

To make it clear how these technologies are involved as building block in IoT, let's take an example of WSN as a building block for IoT (Ray, 2018). Many researchers proposed that WSN with the Internet can work as IoT, but there are lots of issues come into the picture while the researchers integrate WSN with the Internet.

Security and Privacy are the most important issues arise while moving towards IoT. However, security and privacy are always a pre-existing challenge in the field of WSN, but if we connect it with the Internet towards pervasiveness it becomes a more rigorous challenge in IoT. As, it works with the Internet, which inherits the issues involved in terms of IoT, which are as follows:

**Frauds:** We know how fraud has become a trust issue. Fraud is deliberate trickery step perform towards gain advantage. Internet fraud is a type of fraud or deception which makes use of internet and could involve hiding of information for the purpose of tricking victims out of money, property and inheritance just making a cyber trust issue. Juniper research an analyst firm, mobile, online digital market research specialists who provide help in forecasting and consultancy to the technology market. In 2018, Juniper research estimated that frauds in advertising business cost the online advertisers almost US $19 billion worldwide. Further, as Internet is a primary part of IoT and it automatically inherit the pros and cons of Internet frauds. So, IoT automatically suffer from existed frauds.

**Security:** Security is very important in IoT for safe and reliable operation. The authors can say it is the foundational enabler for IoT. Further, the question arises how to implement security at the system, device and network levels. Although, there are a set of network protocols and firewalls provided by various companies, which can manage high-level traffic, passes through the Internet. However, the researchers don't have any solution to protect endpoint devices with a specific task and limited resource to accomplish that particular task. The researchers need to propose a revolutionary novel secure solution to IoT, which can be uniquely applicable to different domains of IoT.

*Figure 2. Application of IoT*

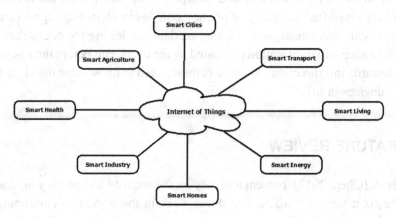

Currently, the researchers don't have any solution that can mitigate every possible cyber threat. Although there are many preexisting solutions addressing previous issues, the researchers can adapt them with the constraint of embedded devices that will increasingly comprise a network of the future.

**Privacy:** IoT is growing rapidly from factory to hospitals, which need better security and privacy that don't leave them vulnerable to data breaches. The nastiest effect of IoT is that end consumers are compromising with their privacy without realizing it because end users are not aware of how the data is being collected and what that data is?

An activist of Electronic Frontier Foundation tweeted about the unsettling similarity of the Samsung Smart TV privacy policy, which warned customers to never discuss personal and sensitive topic near the TV. After this Samsung edit its privacy policy and clarify the data collection process for Smart TV.

The fact is that most of the consumers do not read privacy policies to any new device they buy or applications and if consumers try to read about the policies, most of them are written in a legal language unintelligible to the common consumer. These unintelligible terms include mandatory clauses forcing them to give the right to defend if they are abused by the product, this results compromising the privacy. Further, the industrial service providers can adapt the layered privacy policies, which should be common and creative. The licenses should include the legal code, human-readable, and machine-readable codes. The legal code is the actual policy which is written by the lawyers and interpreted by the judges, human-readable should be a simplified summery of privacy policies in plain language so common consumers can understand, and machine-readable codes are the codes that allows machine to access to information permitted by the users. But before the researchers' moves towards this direction, consumers must asked to know what data is collected by their devices in IoT.

## LITERATURE REVIEW

Chen et al. (Chen, 2017). present a survey on the security and privacy aspect along with the robustness for IoT. IoT systems work in the pervasive environment, so the location of the end devices plays an important role to deliver its facilities. The authors describe the threats related to location, which is categorized according to the global navigation satellite system (GNSS) and non-GNSS solutions. Further, this paper presents certain solutions to provide security and privacy of the positioning systems. It also describes the privacy from both technical and legal aspect with the aims to provide insights into the ideal system that provides robust, secure, and privacy-preserving solution. Khurshid et al. (2018) present a cloud based platform named as Secure-CamFlow, which is concentrated on the data integrity and security aspects while moving from the IoT to cloud. Although, it is not an energy efficient solution, it supports a wide variety of services such as data integrity, privacy and authentication to provide security. It considers flow control policies according to users for privacy preservation. It also uses SHA256, AES, and EI-Gamal cryptosystem as an advance security mechanism. Sivaraman et al. (2018) present the uses of IoT device in home and implication of this scenario from security and privacy point of view. The researchers conclude that it is a complex problem, and there is no single solution available to tackle all the security and privacy aspects of IoT. In this chapter, the authors describe implications for different solution from confidentiality, integrity and authentication, access control, and reflection attacks. Further, the authors try to frame the clause for privacy clauses. In (Dhanjani, 2015) the researchers analyze IoT from different orientation and present the loop-holes in various dimension of IoT. In (Román-Castro, 2018) the authors present the different elements, which is prone to security. It describes all the elements from consumers as well as producer point of view, which include overall security and privacy solution for software and hardware solutions. Recently, Maram et al. (Maram, 2019) propose a secure solution for UNICODE text name as UNICODE data privacy and security (UDPS).

The proposed solution is based on substitution box (S-box) that is dynamic and key dependent solution, although there is a wide verity of solutions existed based on s-box but the dynamic and key-dependent S-box provides the high-level data security. Further, the authors conclude with the maximum avalanche effect for UDPS, which is 73% the highest secure solution available till now. So, there is no "single bullet" solution available, the researcher needs to work on different aspects of IoT from hardware layer to application layer to provide a unified secure solution for IoT. In next section, the authors describe the potential problem at different levels of IoT.

## PROBLEM DESCRIPTION: SECURITY AND PRIVACY ISSUES IN IOT

In the previous section (Introduction), the authors illustrate how security and privacy issues arise while moving towards IoT followed by the existing work approaching to find solutions. In this section, the authors describe what, where, and how security and privacy issues arise. However, the main motive of this chapter is to provide directions towards finding the novel solutions for security and privacy in IoT. In continuation of the previous sections, WSN already has its own security issues and the Internet is an open network which is accessible to anyone from anywhere with its own security and privacy challenges. Moreover, all sister technologies related to IoT have their own security and privacy challenges. While the researchers integrating these sister technologies with the Internet towards IoT, new challenges come into picture with diverse security and privacy issues. The authors describe the problems by considering the integration of technologies, where the actual problem arises and what that problem is?

The researchers need to introduce a security and privacy (Mosenia, 2018) framework for IoT, which is based on the pre-existing approaches for the Internet and the other forms of networks, and pervasive & novel solutions especially for IoT, is a leading research issue.

Figure 3 illustrates the categorization of security and privacy issues of IoT (Kumar, 2016) into three categories, as follows:

### Internet's Own Security and Privacy Issues

The preexisting Internet itself comes with the inherited security and privacy issues. These issues can be solved by using traditional methods for security solutions, e.g., forgery, denial of service attacks, data tempering, and other possible active and passive attacks in Internet. Once Bruce Schneier legitimates person and a cyber security expert says "*Google knows quite a lot about all of us. No one ever lies to a search engine. I used to say that Google knows more about me than my wife does, but that doesn't go far enough. Google knows me even better because Google has perfect memory in a way that people don't.*" Nowadays, Google is the largest search engine in the world and share a very huge data for its users. With all the benefits of the Internet the world has to face a bitter truth i.e., "the exponential decrement in privacy", which is a very serious issue in the world. Today, there are several approaches to jeopardized private information. The authors have categorized main approaches used to jeopardize the user information, as follows.

**Identity Piracy:** It is an act to use the identity of someone else's deliberately to get the private information for financial interest or other benefit with the identity of that particular person. According to India corporate fraud perception survey conducted by Deloitte a multinational company over 30% of companies in India suffers with the identity theft. Further, it provides that Artificial Intelligence, Machine Learning, and Internet of Things are the next generation technologies, which are used in next two years in near future.

**Tracking:** It involves gathering the information about the user to interact with a particular website. Web tracker has the capabilities to collect more information than your browser including personal information like geographical information, IP address, where you came from etc. Each website uses its own cookies. Most of the websites are notifying the users about it, so user need to click "accept", "agree", or "ok" to fetch up the specific webpage to the user. In the era of smart phone privacy issues became bigger. Smart phones are using apps and most of the apps asked to access the complete phone such as storage, emails, contacts, texts etc. and most of the common user are never considers that sharing of such information is a serious issue and a very big privacy threat. Still the technologies are continued to grow and more prone to victimized in some way or simply forced to but stuffs.

Facebook recently announced that they can trace the non-Facebook users also. They are using "Like" button, which is available on almost each and every website. "Like" button contain the cookies and other plug-ins, which are used to store the data of non-user also.

Traditionally, there are a number of solutions available to reduce the threat e.g., Tracker Blocker, Virtual Private Network (VPN), or not to enable unauthorized cookies.

**Protection:** Internet is an open network, which is accessible to anyone from anywhere at any time with these constraints it is not easy to secure this network. Although, there is no idea to make it secure completely, but the researchers can apply several solutions to avoid complete transparency like to deactivate Google tracker, checking the permissions and act accordingly. There are several implications, which can be applied to protect or defense from the existing threats, as follow:

**Computer Access Control:** It includes identifying the user with the help authentication and authorization to approve access control with several constraints. In common word it is simply access approval to computer. So, the user can make a decision weather grant or reject the access based on the prerequisites.

**Application Security:** It basically used to secure application by utilizing software, hardware, and approaches to protect application from external threats. In other words, it is an action taken to improve the security of an application often by finding, fixing, and preventing security vulnerabilities.

Data-centric security: It concentrates on the security of the data rather than the security of application, server, and networks. This relies on the key concepts like discover, manage, protect, and monitor the data flow between the end users.

**Intrusion Detection System (IDS):** IDS is a hardware technology or software application, which is used to monitor the network or system for malicious activity or policy violation. Further, if anyone found in this category, then it will be reported to the administrator, which is responsible to take necessary actions.

**Runtime Application Self-Protection (RASP):** RASP is another technology, which uses runtime instrumentation to detect and block computer attacks by using the information from the backend server. This technology is different from other technologies such as firewalls, which can only detect and blocks attacks and RASP is an enhanced technology that monitor the inputs and block those resource that could allow attack. RASP aims to close the security between application security and network perimeter to provide the complete security solutions.

## Internet's Security Issues Under the Scene of the IoT

There are various solutions for pre-existing security issues in the Internet environment. Although, when the researchers move towards IoT scenario, it introduces some new security challenges. Traditional methods are limited to provide solutions for issues in the Internet. These solutions need to take care about the distinguished characteristics of IoT. Further, the researchers need to introduce the modified security architecture or new architecture, e.g., DNS authentication. In IoT scenario, this will cause the leakage of privacy for object. While the researcher integrates Internet with the technologies this will create new challenges. The research finds the threats with his consideration, as follows:

**User Impersonation and Device Impersonation:** Generally, impersonation refers as an act to be someone for the purpose of fraud. In context of IoT, as it is facilitating its services as an automate system through exchange of the information. So, IoT is more prone to Impersonation as it works autonomously. There are many security measures but the researchers don't have any perfect approach to authenticate user or device and this leads to very serious security threats. Further, the researchers need to introduce some novel approach like certificate mechanism using memory card to authenticate device or users. However, IoT grow exponentially with time along with the large-scale heterogeneity. So, technically it is not possible to cure completely with this issue. The researchers need to find the solution by considering present as well as futuristic point of view.

**Secure Interruption:** It means to the non-availability of services at anytime to anyone from anywhere. In other words, it is an act due to unavailability of information from the resource, which claims to provide the service to its subscribers. Further, the researchers need to introduce novel service selection mechanism to come out of this challenge or the researcher can control using network and access mechanism to outside world.

**Data Diddling:** It is defined as unauthorized and illegal alteration in data, which is used to generate faulty output. IoT produce huge amount of data, intruder either alter the input data or inject malicious function, which process the data to malfunction. As this consider huge amount of data then these crimes are difficult to track. Further, IoT use cloud technologies on large scale to store its data and to provide its services in the form of text, video, image, and sound, which are generated and utilized in different area like educational, health, commercial etc. The communication media, which the system used to transmit the data do not provide secure mechanisms and concern about security this leads to a serious security challenge. The researchers can use available secure mechanism to make it secure like cryptographic, steganographic, and watermarking mechanism to come out of this challenge.

**Worm/Virus Infection:** A warm or virus is a self-replicate piece of code whose main role to infect computers, and will remain active on the system in background. This warm work in hidden mode so it is difficult to identify the warm in the computer. A user realizes the existence of warm, when it consumes all the resource and entire disk space. Further, people can protect our system from this type of code by using updated antivirus software. Although, warm will never enter in the system until user allow them to enter so user should read privacy policy document before using any cookies and malicious software.

**Data Wiretapping:** It originates from electronic surveillance that monitors telephonic and telegraphic communication. Further, with the exponential usage of Internet, wiretapping include to monitor Internet telecommunications. Traditionally, it uses a small plug or a device into telephone but now it is an era of digital communication and all the information is transmitted digitally. So, Internet packet can be tracked through sniffing, which is termed as "packet sniffers". Packet sniffers are behaving as eavesdropper on telecommunication and fetch, what he thinks is relevant. Although, there a law come into picture in 1994 on Wiretapping names as Communication Assistance for Law Enforcement Act (CALEA). However, wiretapping is useful to detect illegal activity happen on Internet.

**OS/Software Vulnerability:** It is considered as a weak property of a system, which can be exfoliated by intruder. To make it possible, an attacker must have at least one interface which is connected to the system weakness. In context of IoT, the operating system should not be vulnerable to make it more secure. The researcher should take care of this entity to make the whole system reliable and robust.

## IoT's Own Security Issues

IoT involves various hardware equipment and other factors. This equipment itself contains various security and privacy threats, which cannot be solved by using traditional solution for Internet. But while moving towards IoT, the researchers need to introduce new innovative solutions by considering various components of IoT. e.g., privacy protection in hardware devices. The number of devices in IoT is very large and these devices are distributed across the globe connectedly. So, any device can communicate to any other device at anytime from anywhere. Smart devices such as domestic appliances, automobile etc. are used to collect and transmit the data across the Internet to facilitate its services to the end users. The end users are unaware of how the data is collected, how much data is collected, and how the data is used to facilitate the services. So, collecting, passing, and sharing of data is very vulnerable to security and privacy attacks.

IoT contains a set heterogeneous device as an end node. They follow different firmware architecture produced by different vendors. Any threat on IoT has a specific intentional purpose and all the threats or attacks are human generated. Although, intent of the attack may vary according to the intruder's target.

- End devices are regulated by human, an intruder may get unauthorized access to the end devices.
- Eavesdropper may get unauthorized access on IoT network devices to get the confidential information.
- IoT device are small in size, power, resource, and processing capabilities. So, these devices cannot allow running complex security protocols, which make it and easy target to intruders.

The authors have categories the hardware attacks that help to illustrate the security issues in IoT, as follows:

## Firmware Modifications

Now a day, most of the modern embedded systems comes with the features of the update firmware. This ability of the embedded system is very useful and provides flexibility to adjust firmware according to the user requirements. But at the same time, it is also vulnerable to attacks. This feature can be exploited to allow intruders to inject malicious firmware modifications into vulnerable embedded systems. In IoT there are verities of devices like switches, routers, and gateways etc. which play an important role in communication or data forwarding. These devices are connected to the computers so any intruder can communicate with these devices and can inject the malicious piece of code into device. Hence, intruder can get all the control over that device and can use it to harm the other users belong to the network.

## Authentication of IoT Devices

IoT devices work autonomously in unattended manner; hence, these devices are easy to be tempered with. Additionally, almost all the commercial and social organizations like social media, retail, etc. collect the user data and analyze it that can be misused.

IoT growing exponentially consequently a large number of IoT devices are attaching, which need authentication to justify the credibility of the device. So, user must authenticate new device by checking their credential using a certificate or key, this leads to reduce the misuse of raw data. Although, several approaches are available to have a secure network. Endpoint security management is one of the important technologies expending in the area of M2M communication and in IoT as well. The researchers need to introduce this type of authentication mechanism for IoT device to maintain the security and privacy.

## Hardware Trojan

It is the baleful circuit for application or modification of the hardware of Integrated Circuit (IC) during designing process or when ICs are fabricated. This results malicious output during runtime. Nowadays, the process of making IC is distributed due to technological requirement and high cost, which is not secure due to dependency of the process. Further, it is almost impossible to detect during the packaging and testing phages because it is possible to inject Trojan circuit at any stage of design, fabrication to manufacturing process.

The researchers use some software tools like Computer Aided Design (CAD) to design on chips, which are developed by some other organization and used in designing process. It may be possible that the software that people are using already have a CAD, which injects Hardware Trojan automatically to the circuit. Hardware Trojan is a very tiny circuit in size so it is almost impossible to find such a tiny circuit in a giant system. Trojan may be detected through reverse engineering process but it destroys the whole IC. Hence, it is impossible to detect Trojan.

The researchers need to make sure the security of the environment and define some rules and policies to design chips securely, which can preserve the policies of third party too. Although, the researchers have proposed other approaches to detect the Trojan in spite of all these unavoidable challenges as Path Delay Testing, Temperature Analysis, Power Based Analysis etc.

## MIDDLEWARE: AN APPROACH FOR SECURITY AND PRIVACY

In this book chapter, the authors present how middleware approach can take care of all the security and privacy challenges at each layer of the IoT layered architecture. Basically, Middleware (Razzaque, 2016) is a software abstraction layer which works above the operating system and below to the application layer. Further, it provides the services to the end user by hiding the complexity of the network. Moreover, it takes care of the privacy of the user as well as of the resources in the architecture.

Middleware distinguishes the user from the actual processing so it can easily take care of the security threats in the architecture by using various traditional approaches (Alam, 2019). Moreover, it reduces the complexity of IoT inherited from traditional networks. Figure 3 illustrates the idea how it distinguish lower layer from the application layer.

However, security and privacy server individually injected in the middleware, which is responsible to handle security and privacy issues of IoT. According to the Oxford English Dictionary (Oxford, 2017) "Software that acts as a bridge between an operating system or database and applications, especially on a network."

These types of systems are used to provide complete solution along with the improved security. If the researchers introduced any system with its security and privacy aspect than middleware is the best practice solution and facilitates designer to develop secure system rapidly. Middleware provide an individual module for security and privacy so the researcher can inject security and privacy measures to the existing solution and can also modify with time, when new security measure comes into picture.

*Figure 3. Categories of IoT security issues*

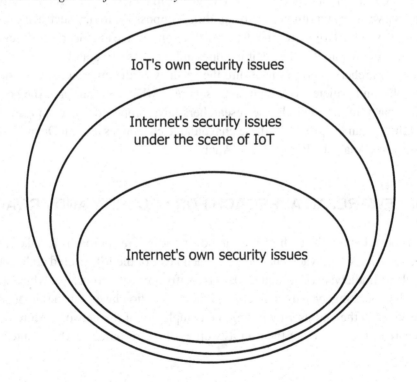

*Figure 4. Basic functional group architecture of middleware*

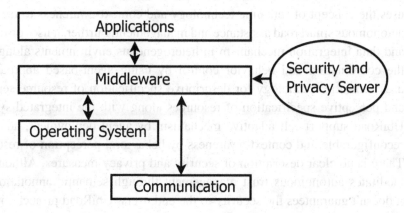

In this book chapter, the authors has reviewed approximately 50 middleware and find the 10 middleware as secured and non-secured middleware. The authors also describe non-secured middleware as these middlewares don't provide security architecture or a minimal description to secure it.

## Non-Secured Middlewares

The authors provide the brief description of the middlewares belong to this category, which are as follows:

**Global Sensor Networks (GSN) (Aberer, 2006):** It provides an architecture, which supports sensor discovery, flexible integration, and data acquisition by distributed query processing. It also facilitates fast deployment and addition of new components dynamically during run time. Further, it uses virtual sensors for resource management, control processing priority and data storage. However, it hides the arbitrary data source behind the virtual abstraction by providing uniform and simple access to heterogeneous components. GSN middleware is adaptive but it didn't have the capability to work autonomously. Although, GSN is widely popular middleware but there is no security constraint according to its architecture. Moreover, the privacy policy is not clearly defined to the end users. Although, in the architecture of GSN there is a place to facilitate access control and integrity checks but it is not sufficient to claim as a secure middleware system.

**UbiRoad (Terziyan, 2010):** It is an example of agent-based middleware, which uses the concept of semantic technology and context-awareness to facilitate autonomous smart road assistance and management. Further, it uses discovery and data integration mechanism in heterogeneous environments along with the coordination and behavior control by using agent-based approach. It uses semantic technology for descriptive specification of resource services and perceptive specification of resources along with the integrated system. UbiRoad support self-adaptive mechanism by agent deployment, adaptive/ reconfigurable and context-awareness by behavioral perception of resource. There is no clear description of security and privacy measures. Although, it facilitates autonomous trust management through semantic annotation but it doesn't guarantees the security to the end users. UbiRoad project with the target of road management, transport management, traffic control, and other use case scenarios so it didn't consider other security and privacy measure.

**UbiSOAP (Caporuscio, 2012):** It is based on service-oriented architecture that provides seamless connectivity in ubiquitous environments. The architecture of UbiSOAP contains necessary function with the abstract components, which is responsible for the interaction between application and resource that provide services through web services. It considers only ubiquitous service without any security constraint.

**WhereX (Puliafito, 2010):** It is event-based middleware for IoT, which hides the complexity of different components. It was mainly designed using RFID for the support of data management. Moreover, initially it was uses to find the related information of the object in the application areas. However, it only considers data management, rather than considering other important factor like security, event handling, etc.

**Mobile Sensor Data Processing Engine (MOSDEN) (Perera, 2014):** It is a plug-in-based middleware to process the collected sensors data for resource constraint devices. The architecture of MOSDEN is based on sensing as a service to the end users, which is built on the top of GSN. This middleware was initially built on Android platform. MOSDEN supports heterogeneity and used as a scalable for end users. Although, MOSDEN provides wide verity of facility but it didn't consider security and privacy aspect for middleware.

## Secured Middlewares

The authors have identified five most popular middleware that describe or implement the satisfactory security parameters. This type of middleware comes into the category of secure middleware.

**Hydra/LinkSmart (Eisenhauer, 2010):** Hydra is an ambient intelligence abstraction layer placed between application layer and operating system. It has a large number of software components and management module for each component (e.g., service manager, device manager etc.). Hydra consists of application and device elements, and each element has semantic layer, service layer, network layer, and security layer to facilitate its services. Moreover, it can work with both existing and new networks of distributed devices. Hydra provides support for heterogeneous devices by using web service interface. It also facilitates both syntactic and semantic interoperability. Additionally, it supports a number of essential requirements of middleware along with dynamic reconfiguration and self-configuration. Further, from the point of view of security and privacy, it uses concept of virtualization along with the web service-based mechanism. It also tries to provide the solution for interoperability and security, but it does not produce a standard solution for globe. Although, these security measures are enough considerable to justify the security of middleware.

**SOCRADES (Guinard, 2010):** It is based on cross-layer service-oriented approach. It is used to extend the reachability from enterprise to real world applications (and vice versa). Moreover, it provides services using device profile web standard (DPWS) and REST to integrate the physical devices into information systems [23]. Further, its services on device facilitate dynamic registration of devices and services, which can be offered by the applications. It facilitates the service discovery offered by real-world devices.

SOCRADES supports services composition but not fully dynamic, it depends on the predefined building block. It provides authentication security by role-based access control mechanism.

**SENSEI (Barnaghi, 2010):** It provides architecture that enable foundation of real-world Internet. It facilitates a number of services for data acquisition, processing of data and provides interaction of services and applications, and context aware management services to real world environments. Further, it consists of three parts: 1. Resource, 2. Support service, 3. Community management. Resources are used for a unified abstraction of devices, support service enable discovery mechanism for devices, dynamic creation and composition of resources for long term interaction, and community management is responsible for handling security and privacy issue related to the applications. SENSEI use semantic ontology modeling for resource management.

**Hermes (Pietzuch, 2004):** Hermes is an event-driven middleware which follows the layered approach. It consists of six layers as: service layer, middleware layer, type and attribute based publish/subscribe layer, type based publish/subscribe layer, overlay routing protocol layer, and network layer [25]. It supports two types of events: attribute based and type-based events. Further, middleware layer divided into two parts: event broker, which provide support to all layer except network layer by providing an external API to connect with other neighbors. Another one is event clients, it follows a component-based model for middleware layer, and event clients are lightweight and language dependent, which provide a connection to local event broker for middleware services. When a new event is introduced, event client generates an event publish request for all the subscribers, then all the interested subscribers can subscribe for the event. Further, Hermes provides support for scalability by using scalable routing algorithms. It also provides support for interoperability and reliability. Although, it has limited support for mobility because of dynamic network topology.

**Sensor Information Networking Architecture (SINA) (Shen, 2001):** SINA is an example of data-driven approach middleware, which was developed at University of Delaware, USA under a research project of Defense department of USA. It provides services by generating query to network and obtain response, and provide commands to distributed sensors to monitor the environmental conditions. It also facilitates the characteristics of scalability and energy efficiency for sensors interaction. It uses SQTL (Sensor Query and Tasking Language) for database access and provides the command (tasks) by programming scripts.

Architecture of SINA is consists of three components hierarchical clustering, attribute-based naming, and location awareness. Hierarchical clustering consists group of nodes on the basis of energy constrain and proximity. These clusters provide levels of clustering on the basis of hierarchical clustering. Further, at each level one cluster head is elected who is responsible for data fusion, filtering, and data aggregation. It

186

uses attribute-based naming to provide an identification of nodes to the final user. However, it may be possible that network is deployed in large areas; in that case location of a node becomes an important factor. SINA uses GPS (Global Positioning System) for location services as primary tool for location awareness, along with beacons by other nodes which don't have GPS system. These nodes can refer to its own position by beacon nodes and relative distance to the beacon. Although, collision may occur because of the large number of responses from different part of network for a query. It may result in response implosion problem. However, SINA use three techniques to overcome from the problem of collision: Sampling operation decides which node takes the part in response, on the basis of response probability. SINA facilitates that all nodes don't need to take part in responses. SINA uses adaptive probability response for that. Next is, Self-orchestrated operation which is used, when there are small number of nodes in network. In this direction, the researchers need to have responses from all the nodes belong to the network which are seeded in different time slot. Last is, diffused computation operation, an aggregation scheme can be defined through SQTL programming script and distributed among all the nodes to avoid information implosion that may occur if all the nodes transfer data directly to front-end.

## PROBLEM DESCRIPTION IN MIDDLEWARE

WSNs are playing an key role in several applications in the field of environment monitoring, agriculture, and smart monitoring, healthcare. Further, WSNs are mainly characterized by many different heterogeneous devices and it consists of many non-proprietary and proprietary solutions. So, it become a big challenge that all the research organization and community would follow same protocol, results difficulty towards achieving the pervasiveness or pervasive integration of sensors with Internet. Although, the people need to move towards closed to standard solutions from proprietary solutions which provides a guarantee solution, while considering all the diverse issues.

This paper provides a comparative survey focused on the main issues related to the evolution of middlewares from WSNs to IoT. The main problems addressed in this paper are associated with middleware are as follows:

- **Interoperation:** Is referred to sharing the information among the diverse devices by using diverse communication protocols, which can be used among the diverse application domains. In WSNs and IoT, interoperability is needed for seamless communication between different domains of application.

- **Lightweight:** Is an important characteristic of middleware. This shows capability to provide same services, either it deployed on power constrained device or in personal computer and cloud system.
- **A Management Capability:** Is an important part of any middleware system, which basically provide support to resource management, device management, memory management etc., related to the applications.
- **Context-Awareness:** Context is used to specify the situation of the entity, it can be a person, place or object relevant to the interaction among users and applications. So, middleware must be context-aware and should work in smart environments.
- **Intelligent Middleware:** Provides an invisible bond that holds critical communications together. It can provide solutions to many security issues related to the IoT applications.

## FUTURE TRENDS AND CONCLUSION

Middleware is a powerful tool to the ease of application development and deployment, which can perform task in complex or critical environments. The middleware framework can facilitate new features and benefits to security and privacy of IoT. This section is helpful to decide the research direction with the goal of security and privacy for IoT along with validation and verification. In this chapter, the authors found some clear gaps. Approximately, 60% proposed solution either has no security constraint or no proper discussion about security of IoT. The authors found potential research gaps in the first section, which provide a wide point of view towards IoT. Further, in next section the authors describe major security and privacy issues and where these issues actual comes for the ease to understand. However, outcome of this chapter is the concept of middleware solution to secure IoT, which is available to wider community of academics and practitioners and application areas. Based on these fact authors believe that there is significant opportunity for contribution to the research community by developing a middleware for IoT along with research gaps, as:

- To develop an architectural model for middleware to enable security and privacy through design (Privacy by design).
- Bringing together common and important characteristics into a unified single middleware, which includes stream processing in cloud, access control through policy, user defined accessibility of data, federated identity (for both user and hardware).

- There is a significant gap to introduce and develop better model by considering the implementation and development threats for the cloud services, which includes discovery, hosting, and usages.
- The researchers need to propose a context-based security model to make it smarter according to reputation of IoT.

In this book chapter, the authors have reviewed 10 middleware according to security and privacy challenges. The authors found some key category as non-secured middleware and secured middleware, which is easy to understand. On the basis of this the authors have identified the future research direction towards security and privacy for IoT.

## REFERENCES

Aberer, K., Hauswirth, M., & Salehi, A. (2006). *The Global Sensor Networks middleware for efficient and flexible deployment and interconnection of sensor networks* (No. REP_WORK).

Alam, T. (2019). Middleware Implementation in Cloud-MANET Mobility Model for Internet of Smart Devices. *arXiv preprint arXiv:1902.09744.*

Althumali, H., & Othman, M. (2018). A survey of random access control techniques for machine-to-machine communications in LTE/LTE-A networks. *IEEE Access: Practical Innovations, Open Solutions, 6,* 74961–74983. doi:10.1109/ACCESS.2018.2883440

Barnaghi, P. (2010). Sensei: An architecture for the real world internet. In *First International Workshop on Semantic Interoperablility for Smart Spaces.*

Caporuscio, M., Raverdy, P. G., & Issarny, V. (2012). ubiSOAP: A service-oriented middleware for ubiquitous networking. *IEEE Transactions on Services Computing, 5*(1), 86–98. doi:10.1109/TSC.2010.60

Chen, L., Thombre, S., Järvinen, K., Lohan, E. S., Alén-Savikko, A., Leppäkoski, H., ... Lindqvist, J. (2017). Robustness, security and privacy in location-based services for future IoT: A survey. *IEEE Access: Practical Innovations, Open Solutions, 5,* 8956–8977. doi:10.1109/ACCESS.2017.2695525

Dhanjani, N. (2015). Abusing the Internet of Things blackouts. *Freakouts, and Stakeouts.* O'Relly Media.

Eisenhauer, M., Rosengren, P., & Antolin, P. (2010). Hydra: A development platform for integrating wireless devices and sensors into ambient intelligence systems. In *The Internet of Things* (pp. 367–373). New York, NY: Springer. doi:10.1007/978-1-4419-1674-7_36

Fujisaki, K. (2019). Evaluation of 13.56 MHz RFID system performance considering communication distance between reader and tag. *Journal of High Speed Networks, 25*(1), 61–71. doi:10.3233/JHS-190603

Guinard, D., Trifa, V., Karnouskos, S., Spiess, P., & Savio, D. (2010). Interacting with the soa-based internet of things: Discovery, query, selection, and on-demand provisioning of web services. *IEEE Transactions on Services Computing, 3*(3), 223–235. doi:10.1109/TSC.2010.3

Katyara, S., Shah, M. A., Zardari, S., Chowdhry, B. S., & Kumar, W. (2017). WSN based smart control and remote field monitoring of Pakistan's irrigation system using SCADA applications. *Wireless Personal Communications, 95*(2), 491–504. doi:10.100711277-016-3905-5

Keramatpour, A., Nikanjam, A., & Ghaffarian, H. (2017). Deployment of wireless intrusion detection systems to provide the most possible coverage in wireless sensor networks without infrastructures. *Wireless Personal Communications, 96*(3), 3965–3978. doi:10.100711277-017-4363-4

Khurshid, A., Khan, A. N., Khan, F. G., Ali, M., Shuja, J., & Khan, A. U. R. (2018). Secure-CamFlow: A device-oriented security model to assist information flow control systems in cloud environments for IoTs. *Concurrency and Computation*, e4729.

Kumar, P., Kunwar, R. S., & Sachan, A. (2016). A survey report on: Security & challenges in internet of things. In *Proc National Conference on ICT & IoT* (pp. 35-39).

Maram, B., Gnanasekar, J. M., Manogaran, G., & Balaanand, M. (2019). Intelligent security algorithm for UNICODE data privacy and security in IOT. *Service Oriented Computing and Applications, 13*(1), 3–15. doi:10.100711761-018-0249-x

Middleware. Oxford English Dictionary. Online. Available at https://en.oxforddictionaries.com/definition/middleware

Mosenia, A. (2018). Addressing security and privacy challenges in internet of things. *arXiv preprint arXiv:1807.06724.*

Perera, C., Jayaraman, P. P., Zaslavsky, A., Christen, P., & Georgakopoulos, D. (2014, January). Mosden: An internet of things middleware for resource constrained mobile devices. In *2014 47th Hawaii International Conference on System Sciences* (pp. 1053-1062). IEEE.

Pietzuch, P. R. (2004). *Hermes: A scalable event-based middleware (No. UCAM-CL-TR-590).* University of Cambridge, Computer Laboratory.

Puliafito, A., Cucinotta, A., Minnolo, A. L., & Zaia, A. (2010). Making the internet of things a reality: The wherex solution. In *The Internet of Things* (pp. 99–108). New York, NY: Springer. doi:10.1007/978-1-4419-1674-7_10

Ray, P. P. (2018). A survey on Internet of Things architectures. *Journal of King Saud University-Computer and Information Sciences, 30*(3), 291–319. doi:10.1016/j.jksuci.2016.10.003

Razzaque, M. A., Milojevic-Jevric, M., Palade, A., & Clarke, S. (2016). Middleware for internet of things: a survey. *IEEE Internet of Things Journal, 3*(1), 70-95.

Román-Castro, R., López, J., & Gritzalis, S. (2018). Evolution and trends in IoT security. *Computer, 51*(7), 16–25. doi:10.1109/MC.2018.3011051

Shen, C. C., Srisathapornphat, C., & Jaikaeo, C. (2001). Sensor information networking architecture and applications. *IEEE Personal Communications, 8*(4), 52-59.

Sivaraman, V., Gharakheili, H. H., Fernandes, C., Clark, N., & Karliychuk, T. (2018). Smart IoT devices in the home: Security and privacy implications. *IEEE Technology and Society Magazine, 37*(2), 71–79. doi:10.1109/MTS.2018.2826079

Terziyan, V., Kaykova, O., & Zhovtobryukh, D. (2010, May). Ubiroad: Semantic middleware for context-aware smart road environments. In *2010 Fifth International Conference on Internet and Web Applications and Services* (pp. 295-302). IEEE. 10.1109/ICIW.2010.50

# Chapter 8

# DDoS Attacks and Defense Mechanisms Using Machine Learning Techniques for SDN

**Rochak Swami**
*National Institute of Technology, Kurukshetra, India*

**Mayank Dave**
ⓘ https://orcid.org/0000-0003-4748-0753
*National Institute of Technology, Kurukshetra, India*

**Virender Ranga**
ⓘ https://orcid.org/0000-0002-2046-8642
*National Institute of Technology, Kurukshetra, India*

## ABSTRACT

*Distributed denial of service (DDoS) attack is one of the most disastrous attacks that compromises the resources and services of the server. DDoS attack makes the services unavailable for its legitimate users by flooding the network with illegitimate traffic. Most commonly, it targets the bandwidth and resources of the server. This chapter discusses various types of DDoS attacks with their behavior. It describes the state-of-the-art of DDoS attacks. An emerging technology named "Software-defined networking" (SDN) has been developed for new generation networks. It has become a trending way of networking. Due to the centralized networking technology, SDN suffers from DDoS attacks. SDN controller manages the functionality of the complete network. Therefore, it is the most vulnerable target of the attackers to be attacked. This work illustrates how DDoS attacks affect the whole working of SDN. The objective of this chapter is also to provide a better understanding of DDoS attacks and how machine learning approaches may be used for detecting DDoS attacks.*

DOI: 10.4018/978-1-7998-0373-7.ch008

## INTRODUCTION

Nowadays, the world has become digitally oriented and full of networking services. Networking is an essential part of our lives because of providing several flexible and easy way of communications. With the increasing growth in advanced network services, chances of cyber-attacks are also growing. There are various attacks that disturb the normal functioning of the networks. One of these attacks is Distributed denial of service (DDoS) attack (Mirkovic et al., 2004). DDoS has become the most frequently used attack for infecting the system's services. It tries to make the services unavailable for normal users by overwhelming it with a huge amount of traffic. DDoS attacks target the system's resources to disrupt the proper functioning of the system's services. Most commonly targeted resources by DDoS attacks are bandwidth, memory, and CPU. These attacks are rapidly growing year by year. As per Arbor's report (Novinson, 2018), DDoS attacks have increased from 1Gbps in 2000 to 100Gbps in 2010, and to more than 800Gbps in 2016 from the perspective of size. One of the biggest DDoS attacks targeted the GitHub in 2018 with a very high rate of traffic. One more such disastrous DDoS attack called "Dyn attack" happened in 2016. It affected the working of many sites such as PayPal, Amazon, GitHub, Netflix, and many more. This attack used a malware named "Mirai" to target these websites. To defend against these vulnerable attacks, more useful research work should be done and efficient intrusion detection system (IDS) should be designed. These IDS systems are very helpful in identifying the attacks in time. Many IDS systems have been developed by researchers and networking companies. A new networking technology "Software-defined networking" (SDN) has also become very famous due to its unique characteristics. Separation of control logic from its data forwarding devices and its centralized global visibility to the entire network topology are two main characteristics of SDN (Nunes et al., 2014). It can become very helpful in DDoS detection using these unique features. SDN can resolve various security issues of conventional as well as trending networking technologies. However, SDN also attracts DDoS attacks due to its centralized controller. DDoS attack targets the SDN controller by sending a large number of malicious packets. By targeting the SDN controller, the whole network can be compromised as a single point of failure. Therefore, efficient defense mechanisms are required to detect the attack in SDN. By overcoming these security issues, SDN serves as a security resolver in a more effective and better way. For these detection mechanisms, machine learning algorithms can be utilized (Michie et al., 1994). Machine learning is widely being used for cyber security. Various machine learning based IDS systems have been proposed by the researchers. They classify the traffic as malicious and normal traffic, which helps to identify the attack. Machine learning based IDS gives better classification results with high accuracy.

This chapter includes a brief literature review of DDoS attacks and its defense using machine learning based IDS. The chapter is organized as follows. Section 2 discusses various types of DDoS attacks and their behavior. A brief overview of working of SDN and the effect of DDoS is given in Section 3. Some major processing modules of an IDS are detailed in Section 4. Section 5 provides some machine learning based DDoS detection approaches. Section 6 provides research challenges in current machine learning solutions. Finally, Section 7 concludes the chapter.

# DISTRIBUTED DENIAL OF SERVICE ATTACKS

DDoS is considered as one of the most serious attacks nowadays. A DDoS is a cyber-attack that attempts to block the online services by overwhelming requests for a time period (Gupta et al., 2009). The target system is forced to slow down, crash or shut down by the flooding of a large number of requests. DDoS attacker keeps the system busy for a certain period of time by forcing the system to serve illegitimate requests consequently denying the services to legitimate customers. One of the most important security principles of the CIA (Confidentiality, Integrity, and Availability) model is availability, which is compromised by DDoS attacks. According to a report (Goodin, 2018), a US based service provider was targeted by a 1.7Tbps attack on 5 March 2018. The attacker spoofed its victim's address and sent a number of packets with ping at a memcached server. The server responded by firing back as much as 50,000 times the data it received. This flooding of traffic was enough to exhaust the server and to deny the services for its legitimate users.

Resources constraints in most of the networking architectures are the main reasons behind the DDoS attacks. These types of attacks mainly consume the communication channel (bandwidth), storage capacity, and CPU processing power, etc.

There are various kinds of DDoS attacks. Different attackers of these attacks have different motives to target the victim. DDoS attacks are rapidly growing in the field of internet, which can be divided into categories as per their target of resource exhaustion.

DDoS attacks can be divided first into two categories as follows:

- **Connection-Based:** It is such type of attack where a connection must be established by the attacker for launching the attack via any standard protocol. These attacks usually affect the web server or applications. Some common examples are TCP and HTTP based attacks.

- **Connectionless:** An attack that does not require a connection/session to be established by the attacker with a victim. It can be launched very easily by transferring the packets to the victim. Some examples of this attack are UDP flooding, ICMP flooding, and many more.

Further, a DDoS attack can be put into the following three categories, which is illustrated in Figure 1.

## Volume-Based Attacks

Volume based attacks are also called as flooding attacks. Flooding attacks are the most commonly used attacks to target a host. In general, 65% of the total reported DDoS attacks are flooding based as per the Arbor's networks (Calyptix, 2015). In this attack, a large volume of packets is forwarded to the target in order to exhaust its bandwidth. These attacks use multiple infected systems, i.e. botnet, zombienet allowing malicious traffic as legitimate. Common types of flooding-based attacks are:

- **UDP Flooding:** UDP is a connectionless protocol and UDP flooding is one of the most common attacks nowadays. The attacker here exhausts the target host with a huge number of UDP packets on the random ports. The target

*Figure 1. Taxonomy of DDoS attacks*

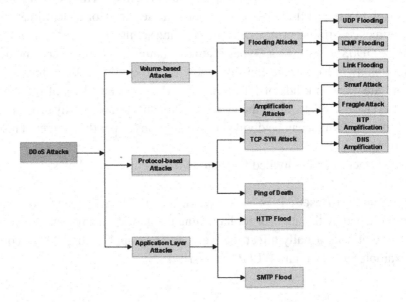

continually checks for application listening at that port. When the target host finds no application at that random port, it sends back an ICMP destination unreachable packet. This attack aims to exhaust the bandwidth and resources of the target host. UDP flooding attack is shown in Figure 2.

- **ICMP Flooding:** It is also known as Ping flooding attack. In ICMP flood, an attacker overwhelms the target host with a huge number of ICMP echo request (ping) packets. Consequently, the target host responds with the same number of echo reply packets that makes both the incoming and outgoing bandwidth of the network exhausted as shown in Figure 3. This attack can become the most successful in case of an attacker having more bandwidth than the victim.
- **Link Flooding Attack:** It is a new type of DDoS attacks that is rapidly growing as a serious threat and can be abbreviated as "LFA". It can break off the connection between legitimate user and victim servers by flooding only a few links. An example of LFA is crossfire attack that can cut off the network connection by flooding only a few links. The origin of LFA is not detectable by any target host because they don't receive any message. They just receive low intensity flows that are not possible to distinguish from legitimate flows.
- **Smurf Attack:** Smurf attack is a network layer DDoS that works somewhat like ICMP flood attack as illustrated in Figure 4. It makes use of a program called smurf malware to enable its execution. Smurf malware generates fake

*Figure 2. UDP flooding*

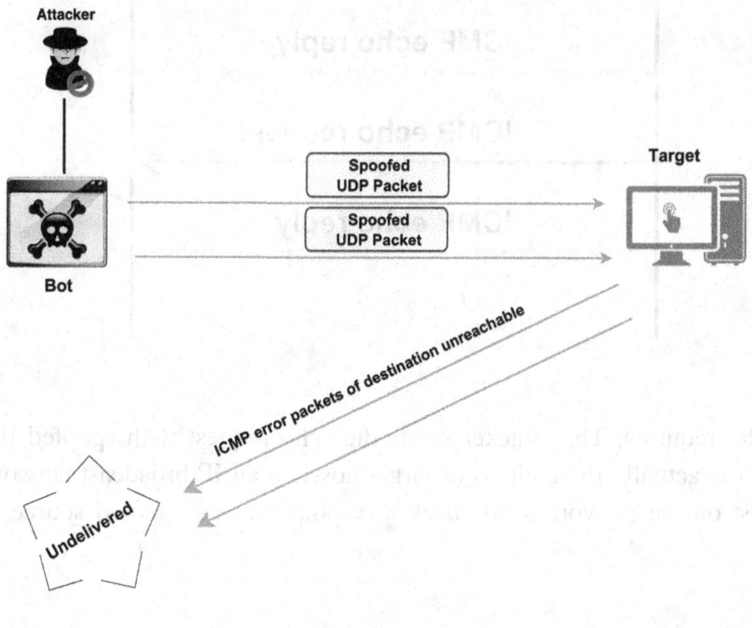

*Figure 3. ICMP flooding*

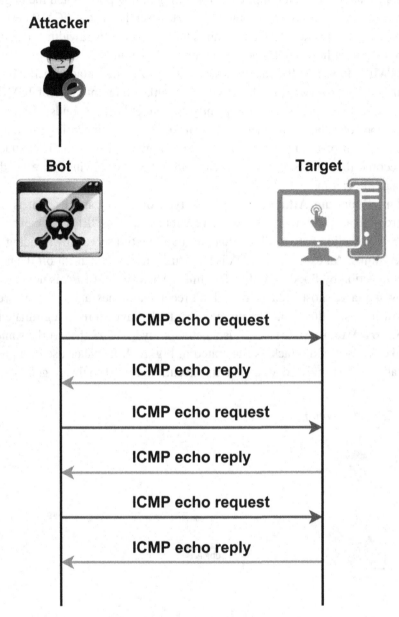

echo requests. The attacker sends the echo request with spoofed IP source that is actually the address of target host, on an IP broadcast network. Each host on the network sends back a response to the spoofed source address.

*Figure 4. Smurf attack*

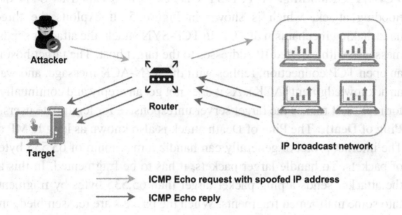

With this flooding of ICMP replies, the target host brings down. It is a type of amplification DDoS attack.

- **NTP Amplification Attack:** It is a reflection-based flooding attack. In Network Time Protocol (NTP) attack, the attacker uses the functionality of NTP servers for flooding on the target host with UDP traffic. The attacker constantly sends "get monlist" command (used for traffic count) to the NTP server with a spoofed IP that is victim's address. The response is sent to the victim server that will be larger than the request. This makes the degradation of the services for legitimate users.

- **DNS Amplification Attack:** It is also a reflection-based flooding attack. In DNS amplification attacks, the attacker exploits the functionality of a publicly accessible DNS server. They send fake DNS query requests with a spoofed IP that is the address of victim server to the DNS server. After receiving several DNS requests, DNS resolver replies back to the victim server with numerous DNS responses. This makes it slow down by overwhelming the victim server and degrades the service for the legitimate users.

## Protocol-Based Attacks

These attacks exploit the network protocols to launch the attack on the target host. They attempt to target the connection state tables in the firewalls, web servers, etc. According to Arbor report (Calyptix, 2015), about 20% of the reported DDoS attacks are protocol-based attacks. Some common examples of protocol based attacks are TCP-SYN flooding and Ping of death.

- **TCP-SYN Flooding:** A TCP-SYN attack is also called as SYN or SYN flooding attack, which is shown in Figure 5. It exploits the three-way handshaking mechanism of TCP. In TCP-SYN attack, the attacker sends SYN messages with spoofed IP addresses to the target host. The target host makes an open TCP connection, replies with the SYN-ACK message, and waits for an acknowledgment (ACK). As it does not get any reply and continually waits for replies, it makes the target server unresponsive for legitimate users.
- **Ping of Death:** The Ping of Death attack is also known as long ICMP attack. The internet protocol generally can handle a maximum of 65,535 bytes size of packets. To handle larger packets, it has to be fragmented. In this attack, the attacker sends a ping packet larger than 65,535 bytes by fragmenting it into some malformed fragments. When the packets are reassembled, an oversized packet is found that leads to memory overflows. The Ping of Death attack can crash or freeze the target server by exploiting a standard protocol. That is why it is classified as a protocol exploitation attack.

## Application Layer Attacks

These attacks exploit the standard application protocols by targeting the online services, web servers, etc. These attacks are considered as the most challenging attacks to be identified or mitigated. As of 2013, 20% of attacks come under application layer attacks (Calyptix, 2015). HTTP flood, Slowloris, and DNS flood are the most common types of application layer attacks. There are some new forms of application layer attacks such as HTML, web browser technology, HTTPS, slow rate attacks, etc. An example of slow rate attack is slow HTTP. They attempt to attack an application

*Figure 5. TCP-SYN attack*

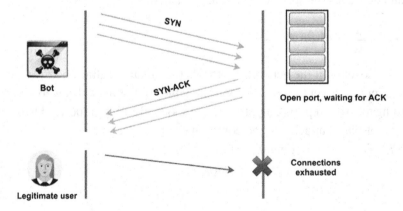

in a way that they appear as actual requests from the legitimate users. According to Arbor network, application layer attacks are the most sophisticated attacks that can be very effective even at a low rate.

- **HTTP Flooding:** HTTP flood is an application layer based volumetric DDoS attack. In this attack, legitimate HTTP GET or HTTP POST requests are exploited to attack a web server or an application. It requires less bandwidth to slow down the target server instead of spoofed, reflection-based techniques. HTTP floods generally utilize a botnet (a group of interconnected computers that are malicious) with malware such as Trojan Horses.
- **SMTP Flooding:** SMTP flood is an application layer-based DDoS attack. In SMTP floods, an attacker exploits the SMTP server by sending a number of anonymous emails. The aim of SMTP attack is to overflow the inbox to slow down the server. The blocking of such emails is not the concern, these emails are automatically identified as spam. There is no way to prevent an email DDoS attack but the chances of being victimized can be reduced using some firewall and security systems.

Recently, a new DDoS attack known as Zero-day attack has appeared. The zero-day attack originates from a security vulnerability unknown to the software developers but known to the attackers. Any patches have not been released yet for this attack. It is named as Zero-day because it occurs before "Day 1", i.e. before the vulnerability becomes publicly known, and considered that the attack occurred on "Day 0".

## SOFTWARE-DEFINED NETWORK

Software-defined network (SDN) has attained a great attention of the researchers in the last few years. The unique property of SDN is the decoupling of the control plane and the data plane (Kreutz et al., 2015; Hakiri et al., 2014). This separation provides agility and flexibility to the network. SDN makes it easy to update the changes required in the network policies as per the user's requirements. These changes have to be made in the control plane only, which reduces the cost of this process. Due to this significant property, SDN is tending to replace the conventional networks in which control and data plane are tightly integrated with each other. A centralized entity "controller" is placed in the control plane that manages and controls all the data plane devices. Therefore, decoupling of control and data plane, and centralized view are the two key characteristics of SDN (Goransson et al., 2016). The design architecture of SDN is described in Figure 6. The complete architecture of SDN is divided into three planes (layers) that are discussed as:

1.  **Data Plane:** It comprises of data forwarding devices, i.e. switches. These SDN switches are responsible for the forwarding of the incoming packets to the destined host. Switches are connected with each other. Each switch has a flow table that contains an entry for the packets. These entries have three fields: rule, action, and counter. On incoming of a packet at the switch, the rule is matched and if the rule doesn't match, the packet is transferred to the controller.
2.  **Control Plane:** It contains a controller, which handles all the data plane devices. The control logic for the forwarding of packets is implemented in the SDN controller. Being a centralized entity, the controller has the capability of global visualization of the network. It decides whether a packet should be forwarded or dropped. The controller instructs the switch in the data plane

*Figure 6. SDN architecture*

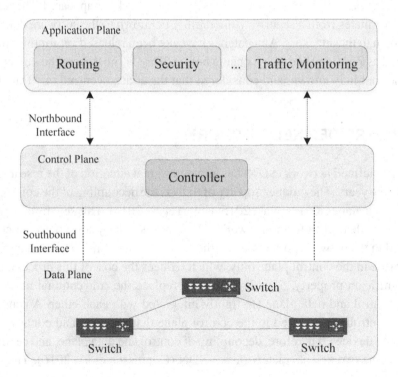

to update the flow table rules timely. In the case of multi-controller scenario, the controllers communicate using east-west bound interfaces. The data plane devices communicate with the controller via a standard southbound protocol named "OpenFlow" (Tourrilhes et al., 2014).

3.  **Application Plane:** This plane consists of various applications required as per the user's and network service's requirements. These applications are deployed over the controller. Some of these applications are network security, traffic monitoring, routing, load balancing, etc. Logics for the data plane devices are implemented in the controller by using these applications. The applications communicate to the controller via the northbound interface.

The packet forwarding between the SDN switches and controller is described in Figure 7. When a new packet come at the switch, it checks its flow table to find out an entry for the incoming packet. A *packet_in* message containing the header of the packet is sent to the controller if the rule doesn't match with the rule placed in the flow table. Then the controller sends back a message called *FlowMod* rule to the SDN switch S1 and S2. *FlowMod* rule instructs the switches to update the corresponding entry of the packet. Both the switches install new flow rules in their respective flow tables. Similarly, if the flow rule gets a match with flow table rule of the incoming packet, the switch takes an action, which is defined in the flow table and forwarded to the next destined switch.

In the case of DDoS attack, the attacker host generates a huge volume of packets and sends to the targeted switch. Since rules of these packets do not match with the flow table rules, so a number of *packet_in* messages are transferred to the controller by the switch. As a result, the controller sends back a number of *FlowMod* rules

*Figure 7. Event-sequence diagram of SDN*

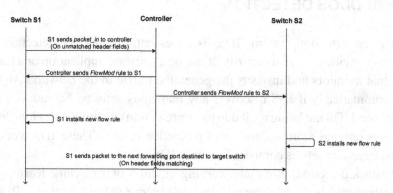

to the switches for updating the rules in flow tables. The controller continuously sends the rule updating messages upto a limit until it gets exhausted. Therefore, flooding of these packets over the channel between switches and controller makes exhausting the controller's resources, switch-controller bandwidth, and switch's flow table. Due to controller and bandwidth exhaustion, legitimate users suffer from the unavailability of the controller's services. This results in dropping out the legitimate packets. It can be observed that DDoS attacks disrupt the functionality of the whole SDN network.

DDoS attacks most commonly target the bandwidth of the communication channel, other resources, i.e. memory, CPU, and power consumption of the system. Based on these targeted resources, the most affected parts of the SDN are listed:

- **Switch:** SDN switches have flow tables to store the packet forwarding rules and its header fields. But the storage capacity of these flow tables is very limited. Therefore, when a number of new packets are sent at a switch, all the storage of switch gets occupied. DDoS attack targets the switch's memory.
- **Switch-Controller Channel:** DDoS attacks send the flood of packets over the communication channel between switch and controller. This flooding of packets exhausts the bandwidth of the channel, which results in dropping the legitimate packets.
- **Controller:** A controller is the main functioning entity of SDN. SDN switches are not able to handle the DDoS attack packets, and then a message request is sent to the controller. Therefore, a large number of messages are transferred to the controller in the case of DDoS. It consumes the controller's resources, i.e. memory and processing that make it unable to serve the services to the legitimate users and the whole working of the network gets collapsed.

## IDS FOR DDOS DETECTION

An intrusion detection system (IDS) is an essential entity for detection of the malicious activities in cyber security. It can be a software application or a hardware device that monitors and inspects the generated traffic in the network. An alert is issued automatically if IDS discovers any malicious activity. Nowadays machine leaning based IDS are being used due to its prediction capability. Machine learning techniques provide more accurate and predictive results. These IDS work better with the large datasets as compared to other IDS.

For attack detection, important working modules of a machine learning based IDS are discussed in this chapter as shown in Figure 8 (Moustafa et al., 2019). This IDS is based on machine learning approaches to identify an attack.

- **Sources of Data:** This module involves collecting the data (network traffic) containing normal and attack traffic. The sources of data can be either real-time data collection or available datasets. For collecting real-time data, the network is generated and captured with some packet analyzing tools. These captured packets are collected in a pcap file, which contains details of packets header, i.e. source IP address, destination IP address, source port, destination port, source/destination MAC address, protocol, timestamp, etc. To access the header fields, the pcap file is converted into a parsed file. Further, some useful features required for attack identification are computed using feature extraction techniques. For developing an effective IDS, there are various off-line datasets available. Most commonly used DDoS attacks datasets are KDD99, NSL-KDD, CAIDA 2017, UNSW-NB15, CICIDS2017, DARPA 2009, ISCX dataset.

- **Preprocessing of Data:** Preprocessing of datasets is an important step in machine learning because real-world data has generally missing values, duplicate values, errors, which may affect the performance of the IDS. Therefore, preprocessing improves the performance of the IDS by resolving these flaws. This step has some functions that are feature creation, feature reduction (feature selection is a special case of feature reduction), conversion of features and normalization. Feature creation is the creation of the features from the captured network packets using network sniffing tools such as Netmate, Scapy, tcpdump, Tcptrace, etc. Using the basic features (source and destination IP addresses), some significant features can also be derived that define network behavior. These derived features are helpful in identifying the attacker and source of the attack. Feature selection is a special case of feature reduction, which includes the selection of some features from an original set of features in the dataset. These features have relevant information only that can be important in the detection of malicious activities. Some most commonly used feature reduction techniques are Principle Analysis Component (PCA),

*Figure 8. Modules of a machine learning based IDS (Moustafa et al., 2017)*

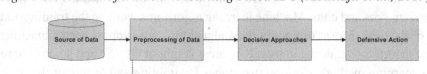

Association Rule Mining (ARM), and Independent Component Analysis (ICA). Datasets have both numerical and text data values. It is required to convert the text data into numerical data. To perform an experiment on the dataset, feature conversion is an important step. Further, data is converted into a normalized form. Feature values are defined into an interval of [0, 1] using normalization function.

- **Decisive Approaches:** It comprises of various approaches to be performed on the preprocessed dataset. These approaches detect malicious activities by classifying them. Machine learning algorithms are used to classify the data into different classes. Different machine learning techniques are support vector machine (SVM), decision tree, artificial neural network (ANN), KNN, etc. It involves two parts in its processing: training and testing of the data. Mostly, binary classification is used which is labeled as "0" and "1". A normal class is indicated with "0" and attack class is represented with "1". In case of multiple classification problems, one class is normal and other classes are classified into a number of classes containing a different type of attack. Based on the training and testing of data, the performance of IDS is evaluated on some indicators i.e. accuracy, detection rate, true negative rate, false positive rate, receiver operating characteristics (ROC) curve, etc.

- **Defensive Action:** Based on the achieved performance results, an appropriate decision is taken and it is defined if the identified class of a packet is normal or attack. An alarm is generated if there is an attack and informed the administrator. Accordingly, the system administrator takes action to stop that malicious activity.

## MACHINE LEARNING BASED DEFENSE SOLUTIONS

Machine learning (Adeli et al., 1994; Tsai et al., 2009) is widely being used in cyber security for the attack detection and classification of the normal and malicious data by the researchers. There is a large variety of machine learning based classification algorithms that are utilized for different purposes. All the algorithms have their own different pros and cons. Machine learning algorithms work on the training and testing of datasets. Most commonly used algorithms are support vector machine (SVM), random forest, decision tree, K-nearest neighbor (KNN) and so on. These machine learning methods work on two stages i.e. training and testing of the data. Performance of these methods is evaluated using some indicators that are accuracy, precision, true positive rate, f-measure, etc. In this section, some proposed solutions by the researchers for DDOS attack detection are discussed.

Kokila et al. (2014) proposed an intrusion detection system against DDoS attacks in SDN. The proposed system used SVM as a classifier to detect the attack. For evaluation, a dataset Darpa 2000 was used. SVM achieved more accurate results with a less false positive rate as compared to some existing methods.

Vetriselvi et al. (2018) proposed an IDS to detect the attacks in SDN based on machine learning and genetic algorithms. The proposed IDS is built up with two modules. One module is responsible for detecting the attack and another module is used for classifying them. The first module is deployed in the SDN switches and the second one is built in the controller. The proposed method of attack detection reduces the load on the controller and also decreases the dependent nature of switches on the controller.

Barki et al. (2016) proposed an approach to detect the DDoS attacks in the SDN environment. The proposed approach is based on two components. The first component is signature IDS and the second one is advanced IDS. Four machine learning algorithms are analyzed for classification of the traffic into normal and malicious traffic. These machine-learning-based algorithms are Naive Bayes, KNN, k-means, and k-medoids. The classification algorithm with higher accuracy is used for the signature IDS. It identifies the normal and malicious behavior of the hosts. A list of malicious hosts is sent to the advanced IDS. The advanced IDS works on the basis of TCP three-way handshaking. The designed IDS is implemented on Ryu controller. Mininet (Yan et al., 2015) is used as an SDN based emulator for the experiment. The results show that Naive Bayes provides a higher detection rate. One disadvantage of the algorithm is that it takes more processing time as compared to using other classification algorithms.

In a research work, Abubakar et al. (2017) presented a machine learning based IDS for securing SDN from attacks. For simulation, a virtual testbed is designed using a star topology with hosts and servers connected to OpenFlow switches. To perform the experiment, OpenDaylight controller (Khattak et al., 2014) and open virtual switches are used with an emulator – Mininet. In the presented work, a signature-based IDS – Snort is used for developing an IDS to detect the attacks. A signature-based IDS is not capable to detect all types of attacks. Therefore, authors also implemented a flow-based IDS using neural network, which has a feature of pattern recognition. It works as an anomaly-based detection. It is combined with signature-based IDS, which can detect attacks that are not detected by signature-based IDS. NSL-KDD dataset is used for the training of the proposed model. Used attack types in this work are DoS, R2L, probe, and U2R. Results show that the proposed model achieves accuracy up to 97%.

da Silva et al. (2016) proposed a framework "ATLANTIC" that uses entropy-based information theory to measure the deviations in the flow table. ATLANTIC utilizes machine learning classification technique to classify the traffic flows. It presents a combination of attack detection, classification, and mitigation tasks in one framework. It consists of two operational phases. The First phase is a lightweight phase, which includes traffic monitoring and attack detection. The second phase is the heavyweight phase, which is responsible for the classification of traffic and mitigation of the attack. In this framework, after capturing all the network traffic flows, features are extracted to perform the classification. Extracted features are *packet_count, byte count,* and *duration.* To classify the traffic flows, SVM is used in the proposed framework. For mitigation, all the malicious packets are blocked and new firewall rules are modified in the flow table. Experiments are performed on Mininet and Floodlight controller. To evaluate the performance of the framework, port scanning and TCP-SYN DDoS attacks are launched. As per the achieved results, all the modules perform well by minimizing the detection overhead.

Ye et al. (2018) designed a DDoS detection model using SVM classification algorithm. For simulation, Mininet and Floodlight controller are used. The proposed attack detection model involves functions performed as part of it: flow statistics collection, features extraction from these flow statistics, and classifying the traffic as malicious using classification algorithm. These flow statistics are extracted from entries in switch's flow table. Six features are used computed for the training of the classifier. In the proposed model, TCP, ICMP, and UDP traffic are used for the evaluation of the experiment. Results show a high detection accuracy.

Yu et al. (2018) designed a model for DDoS attack detection in vehicular networks. The proposed solution is based on SDN capabilities and machine learning. It works on a trigger method which depends on *packet_in* message for timely responding. In the proposed model, features are extracted based on the flow table entries and some useful entropy-based features are also computed. With the help of all these features, a classification model is trained and evaluated. Results show that detection model reduces the attack detection time and has a low false alarm rate.

Chen et al. (2018) proposed a DDoS attack detection system for SDN. In this system, XGBoost classifier is used as an attack detection method in the SDN controller. Attacks, i.e. switch-controller bandwidth congestion and controller's resource exhaustion are discussed in the proposed work. These works include attack traffic generation and packets capturing using simulation tools, and the performance of the detection system is tested. For training the model, an existing dataset is used. As per results, the proposed system achieves accuracy of 98.53%.

Most of the research works have discussed flooding DDoS attacks in SDN. In a research work, Ahmed et al. (2017) presented a DNS query-based DDoS mitigation model. The proposed model involves maintaining traffic statistics and filtering the malicious packets out. It uses a Dirichlet process mixture model to differentiate malicious traffic from normal traffic. Experiments are performed on the real-time data that was collected from network traffic trace files. For classification, features are extracted that are number of transmitted packets, source and destination bytes and duration. Results show that the proposed model performs well for the detection of the DDoS attack as well as for the HTTP and FTP traffic.

Some researchers proposed defense solutions against by developing some intelligence capabilities in the SDN switches. One of these works is presented by Han et al. (2018). Authors proposed a cross-plane DDoS attack defense framework named OverWatch. This framework works on the idea of collaborative intelligence between switches and controller. This framework consists of two modules that are attack detection and reaction. The detection module is responsible for flow monitoring on the SDN switches and an ML classification system on the controller.

These discussed attack detection mechanisms are summarized in Table 1.

## RESEARCH CHALLENGES

This chapter discussed an overview of many existing research works for DDoS detection in SDN, which are based on machine learning. Machine learning is a very effective and significant technique in cyber security for the classification of malicious and normal network traffic. Machine learning based IDS are commonly used in the solving some tasks including regression, prediction and classification But there are also some issues existed that are discussed as:

- Machine learning based IDS provide prediction results with more accuracy. However, in most of the work attack detection approaches are not evaluated for real traffic. All the proposed IDS systems and defense solutions have been measured using already provided datasets, which is not an efficient way. Therefore, there is a need to do more research work in the direction of machine learning based IDS system's development. Researchers should propose such detection and mitigation systems that can be effectively used for the real traffic environment, and that have been evaluated with generated attack traffic.
- There is one more challenge is that a proposed defense solution works for a particular type of DDoS attack. So efficient features should be extracted for the classification of the multiple types of attack and normal traffic. This can be helpful to develop a solution that can be applied for the various DDoS attacks.

*Table 1. Machine learning based detection mechanisms against DDoS*

| Authors | Classification Algorithm | Type of Attacks | Simulation Tools |
|---|---|---|---|
| Kokila et al. (2014) | SVM | BreakIn, DDoS, IPSweep | - |
| Vetriselvi et al. (2018) | SVM | SYN Flood, port scan | Scapy, Floodlight, Mininet |
| Barki et al. (2016) | Naive Bayes, KNN, k-means, and k-medoids | TCP-SYN flood | Ryu, Mininet |
| da Silva (2016) | SVM | Port Scan, TCP-SYN flood | Floodlight, Mininet |
| Ahmed et al. (2017) | DPMM clustering approach | DNS query-based DDoS | - |
| Ye et al. (2018) | SVM | UDP, ICMP, TCP-SYN flood | Floodlight, Mininet |
| Yu et al. (2018) | SVM | UDP, ICMP, TCP-SYN flood | Floodlight, Mininet |
| Abubakar et al. (2017) | Neural network | DoS, Probe, U2R, and R2L attacks | OpenDaylight, Mininet |
| Chen et al. (2018) | XGBoost | UDP, ICMP, TCP-SYN flood | POX, Mininet |
| Han et al. (2018) | Autoencoder based classification | UDP, CIMP, TCP-SYN flood | Ryu, Real-time scenario |

## CONCLUSION

DDoS attacks have become one of the most vulnerable attacks for the networks. They are becoming stronger with the advancement of the network technologies, which make suffered the legitimate users for the network services. Some efficient solutions should be developed for the detection and mitigation of these vulnerable attacks. In this chapter, an overview of various common types of DDoS attacks has been discussed. Different layers of the SDN architecture and impacts of DDoS attacks have also been presented in the chapter. This chapter also includes a discussion of required steps in an IDS that is based on machine learning classification. Finally, authors provided some machine learning based solutions for DDoS detection in SDN. Machine learning based IDS are more efficient in the case of large datasets and provide more accurate classification results. They have the better prediction capability because of having training of the data. There is a requirement of developing efficient machine learning based defense solutions for the real-time network traffic.

# REFERENCES

Abubakar, A., & Pranggono, B. (2017, September). Machine learning based intrusion detection system for software defined networks. In *2017 Seventh International Conference on Emerging Security Technologies (EST)* (pp. 138-143). IEEE. 10.1109/EST.2017.8090413

Adeli, H., & Hung, S. L. (1994). *Machine learning: neural networks, genetic algorithms, and fuzzy systems*. Hoboken, NJ: John Wiley & Sons.

Ahmed, M. E., Kim, H., & Park, M. (2017, October). Mitigating DNS query-based DDoS attacks with machine learning on Software-defined networking. In 2017 IEEE Military Communications Conference (MILCOM) (pp. 11-16). IEEE. doi:10.1109/MILCOM.2017.8170802

Barki, L., Shidling, A., Meti, N., Narayan, D. G., & Mulla, M. M. (2016, September). Detection of distributed denial of service attacks in software defined networks. In *2016 International Conference on Advances in Computing, Communications and Informatics (ICACCI)* (pp. 2576-2581). IEEE. 10.1109/ICACCI.2016.7732445

Calyptix. (2015). DDoS Attacks 101: Types, targets, and motivations. Retrieved from https://www.calyptix.com/top-threats/ddos-attacks-101-types-targets-motivations/

Chen, Z., Jiang, F., Cheng, Y., Gu, X., Liu, W., & Peng, J. (2018, January). Xgboost Classifier for DDoS Attack Detection and Analysis in SDN-Based Cloud. In *2018 IEEE International Conference on Big Data and Smart Computing (BigComp)* (pp. 251-256). IEEE. 10.1109/BigComp.2018.00044

da Silva, A. S., Wickboldt, J. A., Granville, L. Z., & Schaeffer-Filho, A. (2016, April). ATLANTIC: A framework for anomaly traffic detection, classification, and mitigation in SDN. In *Proceedings 2016 IEEE/IFIP Network Operations and Management Symposium (NOMS)* (pp. 27-35). IEEE. 10.1109/NOMS.2016.7502793

Goodin, D. (2018). *US service provider survives the biggest recorded DDoS in history*. Retrieved from https://arstechnica.com/information-technology/2018/03/us-service-provider-survives-the-biggest-recorded-ddos-in-history/

Goransson, P., Black, C., & Culver, T. (2016). *Software defined networks: A comprehensive approach*. Burlington, MA: Morgan Kaufmann.

Gupta, B. B., Joshi, R. C., & Misra, M. (2009). Defending against distributed denial of service attacks: Issues and challenges. *Information Security Journal: A Global Perspective, 18*(5), 224-247.

Hakiri, A., Gokhale, A., Berthou, P., Schmidt, D. C., & Gayraud, T. (2014). Software-defined networking: Challenges and research opportunities for future internet. *Computer Networks, 75*, 453–471. doi:10.1016/j.comnet.2014.10.015

Han, B., Yang, X., Sun, Z., Huang, J., & Su, J. (2018). OverWatch: A cross-plane DDoS attack defense framework with collaborative intelligence in SDN. *Security and Communication Networks, 2018*, 1–15. doi:10.1155/2018/9649643

Khattak, Z. K., Awais, M., & Iqbal, A. (2014, December). Performance evaluation of OpenDaylight SDN controller. In *Proceedings 2014 20th IEEE International Conference on Parallel and Distributed Systems (ICPADS)* (pp. 671-676). IEEE. 10.1109/PADSW.2014.7097868

Kokila, R. T., Selvi, S. T., & Govindarajan, K. (2014, December). DDoS detection and analysis in SDN-based environment using support vector machine classifier. In *Proceedings 2014 Sixth International Conference on Advanced Computing (ICoAC)* (pp. 205-210). IEEE. 10.1109/ICoAC.2014.7229711

Krazit, T. (2018). *What are memcached servers, and why are they being used to launch record-setting DDoS attacks?* Retrieved from https://www.geekwire.com/2018/memcached-servers-used-launch-record-setting-ddos-attacks/

Kreutz, D., Ramos, F. M., Verissimo, P., Rothenberg, C. E., Azodolmolky, S., & Uhlig, S. (2015). Software-defined networking: A comprehensive survey. *Proceedings of the IEEE, 103*(1), 14–76. doi:10.1109/JPROC.2014.2371999

Michie, D., Spiegelhalter, D. J., & Taylor, C. C. (1994). Machine learning. *Neural and Statistical Classification, 13*.

Mirkovic, J., & Reiher, P. (2004). A taxonomy of DDoS attack and DDoS defense mechanisms. *Computer Communication Review, 34*(2), 39–53. doi:10.1145/997150.997156

Moustafa, N., Creech, G., & Slay, J. (2017). Big data analytics for intrusion detection system: Statistical decision-making using finite dirichlet mixture models. In *Data analytics and decision support for cybersecurity* (pp. 127–156). Cham, Switzerland: Springer. doi:10.1007/978-3-319-59439-2_5

Moustafa, N., Hu, J., & Slay, J. (2019). A holistic review of network anomaly detection systems: A comprehensive survey. *Journal of Network and Computer Applications, 128*, 33–55. doi:10.1016/j.jnca.2018.12.006

Novinson, M. (2018). *8 biggest DDoS attacks today and what you can learn from them.* Retrieved from https://www.crn.com/slide-shows/security/8-biggest-ddos-attacks-today-and-what-you-can-learn-from-them

Nunes, B. A. A., Mendonca, M., Nguyen, X. N., Obraczka, K., & Turletti, T. (2014). A survey of software-defined networking: Past, present, and future of programmable networks. *IEEE Communications Surveys and Tutorials, 16*(3), 1617–1634. doi:10.1109/SURV.2014.012214.00180

Tourrilhes, J., Sharma, P., Banerjee, S., & Pettit, J. (2014). SDN and OpenFlow Evolution: A standards perspective. *Computer*, *47*(11), 22–29. doi:10.1109/MC.2014.326

Tsai, C. F., Hsu, Y. F., Lin, C. Y., & Lin, W. Y. (2009). Intrusion detection by machine learning: A review. *Expert Systems with Applications*, *36*(10), 11994–12000. doi:10.1016/j.eswa.2009.05.029

Vetriselvi, V., Shruti, P. S., & Abraham, S. (2018, January). Two-level intrusion detection system in SDN using machine learning. In *Proceedings International Conference on Communications and Cyber Physical Engineering 2018* (pp. 449-461). Springer, Singapore.

Yan, J., & Jin, D. (2015, June). VT-Mininet: Virtual-time-enabled mininet for scalable and accurate software-defined network emulation. In *Proceedings of the 1st ACM SIGCOMM Symposium on Software Defined Networking Research* (p. 27). ACM.

Ye, J., Cheng, X., Zhu, J., Feng, L., & Song, L. (2018). A DDoS attack detection method based on SVM in software defined network. *Security and Communication Networks*.

Yu, Y., Guo, L., Liu, Y., Zheng, J., & Zong, Y. (2018). An efficient SDN-based DDoS attack detection and rapid response platform in vehicular networks. *IEEE Access: Practical Innovations, Open Solutions*, *6*, 44570–44579. doi:10.1109/ACCESS.2018.2854567

# Chapter 9

# Efficient Big Data–Based Storage and Processing Model in Internet of Things for Improving Accuracy Fault Detection in Industrial Processes

**Mamoon Rashid**
*School of Computer Science and Engineering, Lovely Professional University, India*

**Harjeet Singh**
(iD) https://orcid.org/0000-0003-3575-4673
*Department of Computer Science, Mata Gujri College, Fatehgarh Sahib, India*

**Vishal Goyal**
*Department of Computer Science, Punjabi University, India*

**Nazir Ahmad**
*Department of Information Systems, Community College, King Khalid University, Saudi Arabia*

**Neeraj Mogla**
*Nike Inc. Global Headquarters, USA*

## ABSTRACT

*As the lot of data is getting generated and captured in Internet of Things (IoT)—based industrial devices which is real time and unstructured in nature. The IoT technology—based sensors are the effective solution for monitoring these industrial processes in an efficient way. However, the real—time data storage and its processing in IoT applications is still a big challenge. This chapter proposes a new big data pipeline solution for storing and processing IoT sensor data. The proposed big data processing platform uses Apache Flume for efficiently collecting and transferring*

DOI: 10.4018/978-1-7998-0373-7.ch009

*large amounts of IoT data from Cloud—based server into Hadoop Distributed File System for storage of IoT—based sensor data. Apache Storm is to be used for processing this real—time data. Next, the authors propose the use of hybrid prediction model of Density-based spatial clustering of applications with noise (DBSCAN) to remove sensor data outliers and provide better accuracy fault detection in IoT Industrial processes by using Support Vector Machine (SVM) machine learning classification technique.*

## 1. CHAPTER OUTLINE

This chapter is structured around the concepts of efficient storage of sensor IoT based data and its processing in Big Data pipeline. Further the inclusion of novel prediction model will help to improve fault detection in IoT Industrial processes. In today's scenario we have to consider the big source of data generation as well as the plausible suitable platform of such huge data analysis. Therefore, the associated challenges are also included in this chapter.

## 1.1 Introduction to Big Data and IoT

Big Data Analytics along with Internet of Things (IoT) finds its use in areas of Smart Cities, Healthcare, Agriculture and Industrial Automation Units. The challenge of large amount of data generation in IoT devices is fulfilled by Big Data technologies in terms of its storage and processing (Chen et al., 2016). The advanced IoT devices and their applications have given rise to voluminous data in different varieties (Mavromoustakis et al., 2016). On the other side, Big Data technologies have discovered new kind of opportunities for developing IoT based systems (Rashid et al., 2013). Therefore, IoT based Systems and Big Data technologies integration will create new challenges in terms of storage and processing which needs to be addressed by the researchers (Singh et al., 2015).

### 1.1.1 Types of Big Data

The classification of Big Data is mostly given in terms of structure of data. The structure of data depends usually on its organization. Based on this, Big Data is classified into structured, unstructured and semi- structured data (Oussous et al., 2018). These types are explained below:

- **Structured Data:** Structured data is having fixed format and is easily stored, processed and accessed. Structured data is always following particular order as in row and column format and always results into ordered output. This data is easy to process as the format of data is always known in advance. All traditional databases containing data in row column format belong to this category.

- **Unstructured Data:** Unstructured data is usually huge data which is not in organized manner. This kind of data remains usually unknown and poses numerous challenges while processing for valuable insights as output. Moreover, this data is not having any kind of order and is raw in nature. Data in the form of images, audio, video and sensor-based data belong to this category.

- **Semi-Structured Data:** This kind of data usually contains both the forms but remain undefined. Usually this kind of data is not organized inherently at the beginning, but it can be turned into structure form while taking its analysis. Representation of data in terms of XML files belong to this category.

The different types of Big Data are shown in Figure 1.

## 1.1.2 Characteristics of Big Data

The research work on Big Data reveals the following characteristics in terms of its storage and processing (Kapil et al., 2016).

- **Volume:** It is related to amount of data which is getting stored in Big Data file systems. Size of data is one important aspect of Big Data and based on its volume, one can decide whether this data can turn Big Data or not. Volume is the amount of data which is collected and stored in Big Data storage units (Oguntimilehin et al., 2014).

*Figure 1. Different types of big data*

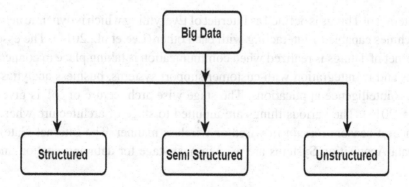

- **Velocity:** It refers to the speed of data which is arriving for its storage or the rate at which data is getting generated from various sources. These days a lot of data is getting generated at high velocities from various sources like Social Media, Sensors, Industrial Units, Weather Forecasting and Mobile devices. The nature of all such remains continuous and come at higher velocities (Kaisler et al., 2013).

- **Variety:** The nature of Big Data remains both structured and unstructured and thus gives it the characteristic of being heterogeneous. Variable data usually consist of data in the form of emails, images, audio and videos. It is always a challenge to analyze the data of these forms (Reddi et al., 2013).

- **Variability:** The nature of Big Data is sometimes decided by the inconsistency of data as well. It is always challenge to process data effectively which is inconsistent in nature (Owais et al., 2016).

### 1.1.3 Advantages of Big Data Processing

- **Improvement in Business Intelligence:** The organizations which are using Big Data platforms access social media like Twitter, Facebook by using various Application Programming Interfaces (API) and get enough insights to fine tune their strategies for the betterment of organization.

- **Improvement in Customer Services:** Big Data technologies resulted in new kind of feedback systems for customers which are far better than traditional systems for getting feedback. The use of Natural Language Processing makes it possible on top of Big Data Platforms for efficient evaluation of Customer responses.

- **Improvement in Operational Efficiency:** Big Data platforms are used for the identification of significant data which is required for its processing and are quite productive in data warehouses.

## 1.1.4 Architecture of IoT

The internet of Things is defined as Internet of Everything which is dynamic network of machines capable of interacting with each other (Lee et al., 2015). The essence of Internet of Things is realized when communication is taking place in connecting devices and its integration with customer support systems, business analytics and business intelligence applications. The stage wise architecture of IoT is given by (Boyes, 2018). The various things are inputted to stage of architecture where the presence of sensors remain in wired or wireless manner. The Internet Gateways and Data Acquisition Systems are used in next stage for data aggregation and its control. The pre-processing and various analytics are performed in next stage for which services in terms Data Centre or cloud is to be used. The whole process is explained in IoT architecture given in Figure 2.

## 1.1.5 Standards of IoT for Industrial Applications

There is no clear line for classifying IoT Standards, however the major standard Protocols in use are based on IOT Data Link Protocols (Salman et al., 2017). Physical layer and MAC layer protocols are mostly used by various IoT standards.

- **IEEE 802.15.4**: This standard is used in MAC layer and specifies source and destination addresses, headers, format of frames, identification in communication between the nodes. Low cost communication and high reliability is enabled in IoT by the use channel hopping and synchronization in terms of time.
- **IEEE 802.11ah:** This kind of standard is used in traditional networking as Wi-Fi for IoT applications. This standard is used for friendly communication of power in sensors and supports lower overhead. This standard covers features of Synchronization Frame, Shorter MAC Frames and Efficient Bidirectional Packet Exchange.
- **WirelessHART:** This standard works on MAC layer and uses time division multiple access. This standard is more secure and reliable than other standards as it uses efficient algorithms for encryption purposes.
- **Z-Wave:** This standard works on MAC layer and was designed specifically for automation of homes. This standard works on Master Slave Configuration where the master sends small messages to slaves and is used for point to point short distances.

*Figure 2. Stage- Wise Architecture of IoT*

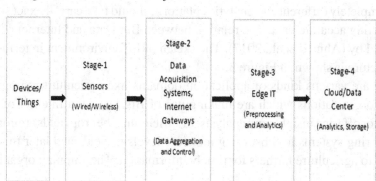

- **ZigBee:** This standard is common one and is used for communication in Health Care Systems and Remote Controls. This standard is meant for medium level communications.
- **DASH7:** This standard is used on MAC layer and is used in RFID devices. This standard supports master slave architecture and is very suitable for IoT applications and is designed for IPv6 addressing.
- **HomePlug:** This standard is MAC based and is used in smart grid applications. The beauty of this standard lies in its power saving mode where it allows nodes to sleep when not required and wakes them only whenever required.

## 2. RELATIONSHIP BETWEEN INTERNET OF THINGS AND BIG DATA

Internet of Things (IoT) is a conversion of range of things into smart objects like refrigerators, vehicles or any electric and electronic gadgets. In IoT, sensors and computer-based chips are used for gathering data in case of those devices which cannot be linked to the internet (Riggins et al., 2015). Whenever this gathered data from smart devices demonstrate volume, velocity and variety, then the role of Big Data comes into picture along with IoT. Usually the data acquired via various sensors remains quite voluminous and carries data in terms both structured and unstructured form. The challenge of velocity in Big Data is the speed of data at which it is getting processed and always shows its presence in terms of IoT based data. Variety is the data in different forms and is one of form in IoT based data. The major challenge in IoT is the way to handle large volumes of data which is getting generated from IoT devices. Big Data Tools are having the capacity to handle this IoT based data with its continuous streaming nature of information. Internet of Things and Big Data is cohesively related as IoT based data is usually raw in nature and it is Big Data Analytics tools which are extracting information from this raw data to get valuable insights to bring smartness in IoT systems. However, the scale for conducting data in IoT is completely different and analytics platform should take care of exact solutions for extracting accurate data. The relation between Big Data and Internet of Things is outlined by (Ahmed et al., 2017). The sketch of IoT environment in terms of IoT Infrastructure is shown in Figure 3.

There are various kinds of application domains like agriculture, shipping and logistics organizations which are making use of Big Data and Internet of Things together for offering insights and analysis. In agriculture, the crop fields are connected to monitoring systems for observing moisture levels in fields and later this data is provided to agriculture farmers for timely information. The shipping organizations

*Figure 3. Broader View of IoT Ecosystem*

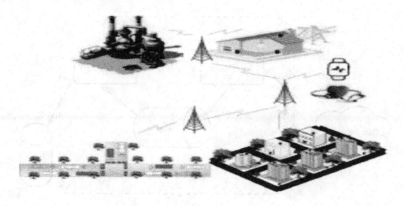

are using sensor data and Big Data Analytics for improving efficiency in terms of delivery of various vehicles to maintain their mileage and speeds.

## 3. ARCHITECTURE OF APACHE FLUME AND STORM

Apache Flume is Big Data tool which is used for ingesting data on Hadoop Distributed File System (HDFS) for storage purpose (Hoffman et al., 2015). This tool is quite beneficial when one needs to collect and transport huge amount of real time streaming data from various sources like social media and sensors in the form of temperature, pressure and humidity. Apache Flume is scalable tool when it comes to large streaming of real data and anytime if the read rate of data from any generator exceeds the rate of writing, this tool provides flow for maintaining the read and write rates steadily (Makeshwar et al., 2015). The basic architecture of Apache Flume is given in Figure 4.

The major components used in Apache Flume architecture are events, agents and clients.

- **Flume Events:** Event is the fundamental unit of data which moves inside Flume pipeline between data source and Hadoop Distributed File System for final storage. Flume Events are data units which are carried out from source to destination with various kinds of headers associated with the data.
- **Flume Agents:** Flume Agents are responsible for the carriage of data from source to sink and receive data from clients. Apache Flume makes use of multiple agents for data transfer purposes until it reaches to final destination. Every Flume Agent internally contains three sub components in the form of Source, Channel and Sink. The source is used for receiving data from the data

*Figure 4. Architecture of Apache Flume for Storage of Real Time Data*

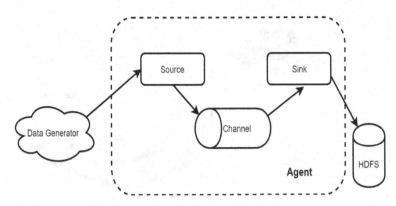

generator and transfers this received data to one of channels in medium. The channel is component which transfers the events received from source and acts as transient buffer until these events are taken by sink. The sink is used for storing the data in file systems like HDFS or HBase.

- **Flume Clients:** Client is the component which generates data in the form of events and then transfers this data to agents for transporting to HDFS environment.

Apache Storm is the real time data processing engine which accepts large volumes of data from various kinds of sources and analyses without storing any actual data (Iqbal et al. 2015). Apache Storm consists of Spouts and Bolts. Initially the data is retrieved from various data sources using Spout which passes this data to various Bolts which perform various operations on this data in the form of analyses and filtration which are later sent for view to end users. The architecture of Apache Storm is given in Figure 5.

Combination of Spouts and Bolts form the topology in Apache Storm and data flow operation takes place in the form of tuples. Data flow in Apache Storm is unidirectional where the data is retrieved by spouts from data sources which are then sent to one or more bolts in one-way direction and remains flowing among blots until the data is finally published to end user (Nivash et al., 2014).

## 4. DATA ANALYTICS FOR IOT USING BIG DATA ANALYTICS

The need of Big Data Platforms and Analytics environment in Internet of Things have increased many folds in last few years and provide valuable benefits and improvement in processes of decision making. Therefore the requirements and demands of Big

*Figure 5. Apache Storm Architecture for Real Time Processing of Data*

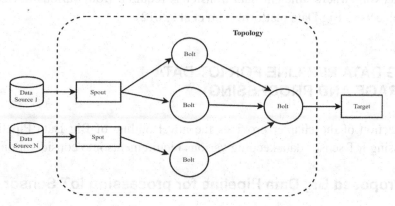

Data Analytics Platforms in Internet of Things have increased for better analytics in its data processing. The inclusion of Big Data Pipelines in Internet of Things have completely changed the way for storing and analyzing the data. The bigger amounts of data generated by various sensor devices can be effectively processed by Big Data Analytics for the extraction of meaningful insights. This section of chapter outlines key requirements required by Internet of Things environment for processing data in Big Data Analytics Platform.

- **Connectivity:** Better connectivity is one of the important requirement in IoT environment for Big Data Analytics on large amounts of machine generated sensor data (Ahmed, E. et al. 2017). Reliable connectivity is a way for connecting infrastructures with high performance with various kinds of objects for enabling services of Internet of Things.
- **Streaming Analytics:** This kind of data analytics is another key requirement in IoT environments and deals with real time data in motion. Data streams which are real time in nature are analyzed for the detection of critical situations. Big Data Analytics platforms require data on the fly and process it in the form of data streams (Tönjes et al., 2014).
- **Storage:** Another key requirement in IoT based Big Data Platforms is the storage of huge amounts of data generated by various IoT objects on some commodity storage units with low latency factors in its analytics. M2M communication protocols are widely used in most Internet of Things services for handling large streams and provide benefits of cloud systems in terms of distributed storage (Suciu et al., 2015).
- **Quality of Services:** The requirement in IoT based Big Data Platforms is the quality of Services in mobile devices and IoT sensors in terms of resource management. The Quality of Service must be quite efficient in IoT based

network where efficient data transfer is required from various devices and objects to Big Data platforms (Jin et al., 2012).

## 5. BIG DATA PIPELINE FOR IOT DATA STORAGE AND PROCESSING

This section of the chapter provides the novel outline of Big Data Pipeline for processing IoT sensor data keeping various requirements into consideration.

## 5.1 Proposed Big Data Pipeline for processing IoT Sensor Data

The proposed model takes care of storage of IoT based sensor data and then processes it on the Big Data Real Time Engine and later uses hybrid prediction model for detecting faults in industrial assemblies. The hybrid prediction model for fault detection is discussed in the next section. The idea is to connect IoT based sensors to the industrial assembly pipelines for sensing the data. The temperature, accelerometer and humidity sensors are used for capturing data from assemblies. The IoT sensor-based data is transmitted to the cloud server which is later stored in Big Data storage unit of Hadoop Distributed File System in terms of large volumes. The outliers in stored sensor data are filtered with the help of clustering method of Density-based spatial clustering of applications with noise (DBSCAN) where from the fault predictions are taken by applying machine learning classification technique of Support Vector Machine (SVM). The data in HDFS is processed on real time basis with the help of Apache Storm. The structure of proposed pipeline for storage and processing of IoT data is given in Figure 6.

*Figure 6. Proposed Big Data Pipeline for Storage and Processing IoT Data*

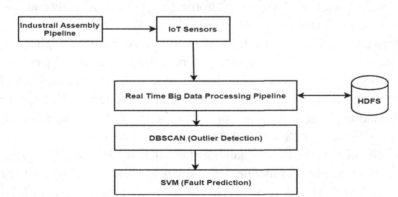

## 5.2 Hybrid Prediction Model for Fault Detection

Outlier detection is done with the help of Density-based spatial clustering of applications with noise (DBSCAN) for given dataset (Ester et al., 1996). After the removal of outliers from sensor data, Support Vector Machine (SVM) is used for the prediction of faults. A classification model is trained in terms of SVM classifier for predicting faults in terms of events are taking place normally or abnormally due to halts.

### 5.2.1 Step Procedure of DBSCAN Algorithm

For given dataset of T, DBSCAN will work on two parameters. $\varepsilon$ (eps) and minPts. $\varepsilon$ is the measure of distance between two assumed points of 'p' and 'q' for checking the density reachability from neighbors. minPts are the minimum number of points which are required to form the cluster.

Step 1: Any point (P) is assumed as starting point which has not been visited before.

Step 2: Select neighbors of this arbitrary starting point (P) on the basis of distance with $\varepsilon$.

Step 3: If density of neighbors is achieved for the point (P), then it is marked as visited and clustering begins. Otherwise it is marked as noise.

Step 4: If P is in cluster, then $\varepsilon$ in its neighborhood is in cluster as well.

Step 5: Repeat Step 2 for all $\varepsilon$ neighborhood points until all points in cluster are taken.

Step 6: New point which is unvisited and marked as clustering point or noise.

Step 7: Repeat Step 2 to Step 6 until all points are visited and marked.

### 5.2.2 Features of DBSCAN Algorithm

1.  DBSCAN is a clustering algorithm which is not relying on specifying the number of clusters for a given data.
2.  DBSCAN algorithm is very robust for outliers and have tendency to filter them out.
3.  DBSCAN algorithm is quite useful in finding patterns and predicting trends where data is complex and hard to find manually.
4.  DBSCAN algorithm has a tendency to produce variable number of clusters depending on the size of input.

Support Vector Machine (SVM) is one of the popular classification machine learning algorithm which effectively segregates classes within a hyper-plane (Evgeniou et al., 2001). Hyper-plane in n-dimensional plane is n-1 dimensional subsets in space which divides it into parts that are disconnected. SVM classifier makes the prediction decisions of classifying with a wider boundary between the classes.

The data from IoT sensors is stored on Hadoop Distributed File System (HDFS) with faulty data as well. The collected dataset is labelled with attributes of normal and faulty for classification purposes. Once the outliers in data DBSCAN, the refined data is finally inputted to SVM classifier for the detection of faults in data. The achieved results are compared with Logistic Regression, Naïve Bayes and Random Forest. The proposed model of DBSCAN+SVM showed increase in detection rate of faults in comparison to existing state-of-art. The performance of hybrid prediction model in terms of various performance measures is shown in Table 1.

Proposed model is evaluated for performance in terms Logistic Regression, Naïve Bayes, Random Forest and SVM classifiers and the results achieved for hybrid prediction model of DBSCAN +SVM are better than other classifiers. The comparison performance accuracy is plotted in Figure 7.

*Table 1. Comparison of prediction models for fault detections*

| Model | Accuracy (%) | Precision | Recall |
|---|---|---|---|
| Logistic Regression | 96.91 | 0.979 | 0.978 |
| Naïve Bayes | 93.50 | 0.939 | 0.936 |
| Random Forest | 98.05 | 0.985 | 0.984 |
| SVM | 97.35 | 0.928 | 0.926 |
| DBSCAN+ SVM | 98.85 | 0.990 | 0.987 |

*Figure 7. Comparison of accuracy for various classifiers*

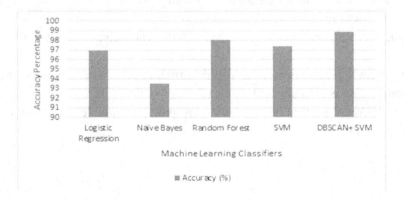

Proposed model is evaluated for precision in terms Logistic Regression, Naïve Bayes, Random Forest and SVM classifiers and the results achieved for hybrid prediction model of DBSCAN +SVM are better than other classifiers. The comparison performance measure for precision is plotted in Figure 8.

Proposed model is evaluated for recall in terms Logistic Regression, Naïve Bayes, Random Forest and SVM classifiers and the results achieved for hybrid prediction model of DBSCAN +SVM are better than other classifiers. The comparison performance measure in terms of recall is plotted in Figure 9.

*Figure 8. Comparison of precision for various classifiers*

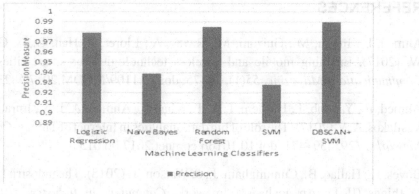

*Figure 9. Comparison of recall for various classifiers*

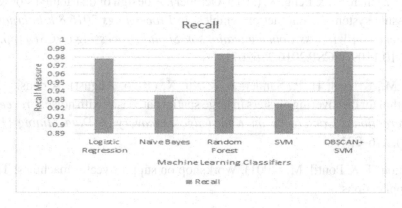

## 6. CONCLUSIONS AND FUTURE DIRECTIONS

In this research, the authors have tried to solve the challenge of real time processing of sensor-based data and fault detection of industrial assembly data. The IoT based sensor data is stored on Hadoop Distributed File System (HDFS) which is processed with Apache Storm on Big Data processing pipeline. The outliers in stored data were removed with the help of Density-based spatial clustering of applications with noise (DBSCAN) and then faults in industrial processes were detected using Support Vector Means classifier. The results suggest that this hybrid prediction model is scalable for data processing of IoT based sensor data and detections in faults are much better than traditional models. However, there is still room of improvement for this model in terms of its training for multi-fault system where the data to be trained is of complex type.

## REFERENCES

Ahmed, E., Imran, M., Guizani, M., Rayes, A., Lloret, J., Han, G., & Guibene, W. (2017). Enabling mobile and wireless technologies for smart cities. *IEEE Communications Magazine*, *55*(1), 74–75. doi:10.1109/MCOM.2017.7823341

Ahmed, E., Yaqoob, I., Hashem, I. A. T., Khan, I., Ahmed, A. I. A., Imran, M., & Vasilakos, A. V. (2017). The role of big data analytics in Internet of Things. *Computer Networks*, *129*, 459–471. doi:10.1016/j.comnet.2017.06.013

Boyes, H., Hallaq, B., Cunningham, J., & Watson, T. (2018). The industrial internet of things (IIoT): An analysis framework. *Computers in Industry*, *101*, 1–12. doi:10.1016/j.compind.2018.04.015

Chen, Z., Chen, S., & Feng, X. (2016, October). A design of distributed storage and processing system for internet of vehicles. In *Proceedings 2016 8th International Conference on Wireless Communications & Signal Processing (WCSP)* (pp. 1-5). IEEE. 10.1109/WCSP.2016.7752671

Ester, M., Kriegel, H. P., Sander, J., & Xu, X. (1996, August). A density-based algorithm for discovering clusters in large spatial databases with noise. *Proceedings of International Conference on Knowledge Discovery & Data Mining (KDD)*, *96*(34), 226–231.

Evgeniou, T. & Pontil, M. (2001). Workshop on support vector machines: Theory and applications.

Hoffman, S. (2015). Apache flume: Distributed log collection for hadoop. Birmingham, UK: Packt Publishing.

Hsu, C. W., Chang, C. C., & Lin, C. J. (2003). A practical guide to support vector classification.

Iqbal, M. H., & Soomro, T. R. (2015). Big data analysis: Apache storm perspective. *International Journal of Computer Trends and Technology*, *19*(1), 9–14. doi:10.14445/22312803/IJCTT-V19P103

Jin, J., Gubbi, J., Luo, T., & Palaniswami, M. (2012, October). Network architecture and QoS issues in the internet of things for a smart city. In *2012 International Symposium on Communications and Information Technologies (ISCIT)*, (pp. 956-961). IEEE. 10.1109/ISCIT.2012.6381043

Kaisler, S., Armour, F., Espinosa, J. A., & Money, W. (2013, January). Big data: Issues and challenges moving forward. In *Proceedings 2013 46th Hawaii International Conference on System Sciences* (pp. 995-1004). IEEE.

Kapil, G., Agrawal, A., & Khan, R. A. (2016, October). A study of big data characteristics. In *Proceedings 2016 International Conference on Communication and Electronics Systems (ICCES)* (pp. 1-4). IEEE.

Lee, I., & Lee, K. (2015). The Internet of Things (IoT): Applications, investments, and challenges for enterprises. *Business Horizons*, *58*(4), 431–440. doi:10.1016/j.bushor.2015.03.008

Makeshwar, P. B., Kalra, A., Rajput, N. S., & Singh, K. P. (2015, February). Computational scalability with Apache Flume and Mahout for large scale round the clock analysis of sensor network data. In *Proceedings 2015 National Conference on Recent Advances in Electronics & Computer Engineering (RAECE)* (pp. 306-311). IEEE. 10.1109/RAECE.2015.7510212

Mavromoustakis, C. X., Mastorakis, G., & Batalla, J. M. (Eds.). (2016). *Internet of Things (IoT) in 5G mobile technologies* (Vol. 8). Springer. doi:10.1007/978-3-319-30913-2

Nivash, J. P., Raj, E. D., Babu, L. D., Nirmala, M., & Kumar, V. M. (2014, July). Analysis on enhancing storm to efficiently process big data in real time. In *Proceedings Fifth International Conference on Computing, Communications and Networking Technologies (ICCCNT)* (pp. 1-5). IEEE. 10.1109/ICCCNT.2014.7093076

Oguntimilehin, A., & Ademola, E. O. (2014). A review of big data management, benefits and challenges. *A Review of Big Data Management, Benefits and Challenges, 5*(6), 1–7.

Oussous, A., Benjelloun, F. Z., Lahcen, A. A., & Belfkih, S. (2018). Big Data technologies: A survey. *Journal of King Saud University-Computer and Information Sciences, 30*(4), 431–448. doi:10.1016/j.jksuci.2017.06.001

Owais, S. S., & Hussein, N. S. (2016). Extract five categories CPIVW from the 9V's characteristics of the Big Data. *International Journal of Advanced Computer Science and Applications, 7*(3), 254–258.

Rashid, M., & Chawla, R. (2013). Securing data storage by extending role-based access control. *International Journal of Cloud Applications and Computing, 3*(4), 28–37. doi:10.4018/ijcac.2013100103

Reddi, K. K., & Indira, D. (2013). Different technique to transfer Big Data: Survey. *Int. Journal of Engineering Research and Applications, 3*(6), 708–711.

Riggins, F. J. & Wamba, S. F. (2015, January). Research directions on the adoption, usage, and impact of the internet of things through the use of big data analytics. In *2015 48th Hawaii International Conference on System Sciences (HICSS),* (pp. 1531-1540). IEEE. 10.1109/HICSS.2015.186

Salman, T. & Jain, R. (2017). A survey of protocols and standards for internet of things. *Advanced Computing and Communications, 1*(1).

Singh, P., & Rashid, E. (2015). Smart home automation deployment on third party cloud using internet of things. *Journal of Bioinformatics and Intelligent Control, 4*(1), 31–34. doi:10.1166/jbic.2015.1113

Suciu, G., Suciu, V., Martian, A., Craciunescu, R., Vulpe, A., Marcu, I., & Fratu, O. (2015). Big data, internet of things and cloud convergence–an architecture for secure e-health applications. *Journal of Medical Systems, 39*(11), 141. doi:10.100710916-015-0327-y PMID:26345453

Tönjes, R., Barnaghi, P., Ali, M., Mileo, A., Hauswirth, M., Ganz, F., & Puiu, D. (2014, June). Real time IoT stream processing and large-scale data analytics for smart city applications. In *poster session, European Conference on Networks and Communications.* sn.

# Chapter 10
# Blockchain Technology Integration in IoT and Applications

**Bhanu Chander**

 https://orcid.org/0000-0003-0057-7662
*Pondicherry University, India*

## ABSTRACT

*The Internet of Things (IoT) pictures an entire connected world, where things or devices are proficient to exchange a few measured data words and interrelate with additional things. This turns for a feasible digital demonstration of the existent world. Nonetheless, nearly all IoT things are simple to mistreat or compromise. Moreover, IoT devices are restricted in computation, power, and storage, so they are more vulnerable to bugs and attacks than endpoint devices like smartphones, tablets, and computers. Blockchain has remarkable interest from academics and industry because of its salient features including reduced dependencies on third parties, cryptographic security, immutability, decentralized nature, distributed nature, and anonymity. In the current scenario, blockchain with its features provides an anonymous framework for IoT. This chapter produces comprehensive knowledge of IoTs, Blockchain knowledge, security issues, Blockchain integration with IoT (BIoT), consensus, mining, message validation mechanisms, challenges, a solution, and future directions.*

DOI: 10.4018/978-1-7998-0373-7.ch010

## INTRODUCTION

The standard of Internet of Things (IoT) positioned solid way for a world where several days by day things interrelated and cooperate with the intrinsic atmosphere in order to gather knowledge for routine secure tasks. IoT is expanding its appliances at fast face some reports already predicated that IoT connectable devices (smartphones, laptops, tablets, etc.) will grow to 30 billion by 2020-22 moreover machine to machine connectivity also grow from 800 million to 3 billion, all of them allied with wide-ranging applications similar to healthcare management, transportation, smart society, justification, and automated robots. In support of trustworthy with broad network intensification, it is compulsory to fabricate appropriate IoT protocols, architectures those can provide IoT services. At present IoT hang on centralized server-client prototype through Internet, but present prototype may not work in future because of rapidly growing devices usages, to overcome this fresh paradigm/ prototype to be planned. Such as exposure entails, stuck between supplementary things, vigor against attacks, faultless authentication, data privacy and security, straightforward deployment and self-maintenance. These above-mentioned features obtain as a result of Blockchain (BC) the skill back-of-the Bitcoin crypto-currency scheme, studied to fascinating as well as critical for ensuring security and privacy for diverse applications in many domains-including the internet of things (Tiago et al., 2018; Mahadi et al., 2018; Ana Reyna et al., 2018).

Abstractly blockchain technology is a distributed database that holds proceedings of transactions which were collective among participated entities; every transaction is well-established by agreement of greater part of a group member, creating forged transactions unable to pass shared conformation. If one time a blockchain shaped and when established, it cannot be altered or modified. Blockchain technologies proficient to track, synchronize complete transactions and accumulate records/data from an outsized quantity of devices which make possible construction of applications that are entailing no central based cloud.

The hurried development of blockchain technology and blockchain involved appliances have begun to revolutionize the digital world's finances as well as financial services. At present, the appliances of blockchain raise from a financial transaction or insurance claim to issues in share trades and corporate bonds. In presumption, any-person any-place can utilize blockchain technology to broadcast information/ data steadfastly. Blockchain technology is kind of distributed ledger knowledge that efficiently removes the central data point rather than most commonly used supply chains data structures. Here, distributed ledger knowledge/technology is the heart of blockchain mechanism, which offers validation method via a network of computers that make possible peer to peer transactions devoid-of the requirement for mediator/ centralized authority to inform and maintain the information generated at the time of transactions. Each and every transaction in blockchain technology validated through a group of validated transaction operations, then after added as a new block to a previously existed chain of transaction operations, because of this reason it named as Block-chain. Once a transaction fixed, added to chain transaction it cannot be altered, deleted or modified. Notably there are two most successful blockchain networks available – public or permissionless blockchain networks – Each user can individually access and perform much like open-source network, permission blockchain networks – specific individuals or organizations use to conduct transaction operations (Deepak et al., 2018; Ana Reyna et al., 2018; Tiago et al., 2018).

Internet of Things (IoT) is one of the emerging topics in recent time in terms of technical, social and financial consequences. From the past decades, there is a significant development in the fields of wireless communication technology, information, and communication systems, industrial designs, and electromechanical systems encourage to progress new technology named as the Internet of things. The most important intention IoTs is to connect all or any devices to the internet or other connected devices. Internet of things is collection network of home appliances, physical devices, vehicular networks and other devices fixed with sensors, electronics, actuators moreover network connectivity which make-possibility for mentioned objects to gather/accumulate and exchange information/data. IoT works as a massive network of interconnected things, people those can collect and share resources about the way they are utilized and about the surrounding environment to them. Here each and everything/device typically identified with its corresponding computing system, however, is able to interoperate within the existing internet infrastructure (Deepak et al., 2018; Ana Reyna et al., 2018; Clemence et al., 2018).

## TAXONOMY OF IOT SECURITY ISSUES

Internet of Things prototype covers, integrates numerous amounts of device collections from undersized implanted processing chips to outsized-end servers, at the same level IoT have to deal with defense troubles at dissimilar levels (Minhaj et al., 2018).

## Physical and Data Link Layer Attacks

Both physical, as well as data link layers, execute key role to accumulate and organizing information at the starting phases of IoT services. Security at this stage concerned as the first level or low-level of security. Here some security issues of physical layers of hardware and data link layer of communication mentioned below.

1.  **Insecure Physical Surface:** IoT consists of a wide variety of hardware devices, various physical factors composite severe exposures to accurate functioning of IoT devices. Unhealthy physical defense, software availability during physical interfaces, testing tools can probably demoralize nodes in the surrounding network.

2.  **Insecure Initialization:** Proper initialization of IoT at preliminary level makes possibilities for appropriate functionality of complete organization without infringing as well as the distraction of network services. Secured physical level communications remove unauthorized receivers.

3.  **Jamming Attacks:** Adversaries with a jamming attack on wireless-based sensor devices within IoT aim to decline communication arrangement through generating radio regularity signals without a definite procedure. These types of radio obstructions severely impact the network operations and disturb the legitimate entities sending and receiving procedures.

4.  **Sleep Deprivation Attack:** Wireless sensor devices utilized in IoT are energy-constrained which makes them vulnerable to sleep deprivation attacks to stay away from the whole network. It will cause issues like battery exhaustion if there are a number of tasks to be carried out on a single sensor node.

5.  **Sybil Attacks:** Adversaries with Sybil attacks generate fake identities to degrade IoT functionalities. And also starts aiming to decrease node or network resources power.

## SECURITY SOLUTIONS FOR PHYSICAL AND DATA LINK LAYER ATTACKS

Jamming associated attacks for wireless sensor networks follow-on messages collision or overflows communication channel. (Young et al., 2011) detects jamming attacks by calculating the strength which is then used for extorting noised bases signals, after that evaluate with modified threshold values. (Xu et al., 2005) proposed jamming attack detection with packet delivery ratio. Cryptographic based error-correcting code presented by (Mounir et al., 2003). (Pecorella et al., 2016) designed a framework for secure physical layer communication where the smallest data rate is constituted among transfer and receiver to make sure the absences of adversaries. (Demirbas et al., 2006) designed a method against Sybil attacks, in their method detector nodes deployed to calculate the sender location at the time of message communication. One more message announcement with similar sender position but special sender personality indirect as Sybil attack. (Li et al., 2006) utilizes radio sign strength dimensions for MAC addresses for spot spoofing stacks. (Xiao et al., 2009) come with a new approach where the amount of uniqueness and extra parameters associated with channel evaluation, spot Sybil attacks.

## Routing, Transport and Session Security Issues

Here some of the security issues mainly focused on communication, routing and session management of IoT devices are described below (Minhaj et al., 2018):

1. **Routing Attack:** The IPv6 routing procedures for LoWPAN are defenseless against numerous attacks through compromised nodes. The main reason for attacks is to decrease the resources of network devices.
2. **Sinkhole Attacks:** With the help of the sinkhole attack, attackers make sensor node more powerful than its adjoin nodes and try to transmit information through the attacker node then performs malicious activities on the network. Moreover, it transmits all packets of the selected region into the malicious node.
3. **Wormhole Attack:** In the wormhole attack, the attacker creates a short concentration link involving two segments of the network over that attacker replays network messages. This kind of attacks has rigorous implications include eavesdrop of service, privacy breach.
4. **Replay or Duplication Attacks:** Attacker intensions is interrupted message and try to retransmit the same message in upcoming time intervals so that bandwidth of the sensor network decreases. A modernization of packet portion fields at the 6LoWPAN layer could affect into the exhaustion of possessions, buffer run over as well as rebooting devices.
5. **End to End Security:** Transport layer provides a secure mechanism from end-to-end, so data from the initiator node to the receiver node in a consistent approach. To achieve the above-mentioned approach a well-dressed encrypted authentic mechanism to be applied that guarantees secure message communication and not breach privacy while working in an insecure channel.

6. **Buffer Reservation Attack:** In IoT, servicing node needs to preserve buffer gap used for a rebuild of arriving packets; here an invader might be utilizing it by transport imperfect packet fragments. This will lead to denial-of-service where legitimated fragments denied owing to the space engaged through imperfect packets send by the invader.

7. **Session Establishment:** The main intention is to take control of the session on transport layer with malicious nodes results in a denial-of-service attack. Moreover, the attacker impersonates the attacked node to prolong the session among two nodes.

8. **Authentication:** IoT is a collection of numerous devices and abusers, so key management process exploited to authenticate one with other devices or with abusers. Any kind of miss connection in security at network layer communication can represent the network to a great number of attacks. Convolution cryptographic mechanism ensures secure communication but concedes more energy from sensor nodes which have limited resource constraints.

9. **Privacy on Cloud-Based IoT:** Location privacy violation may deploy on cloud interrupt broadcasting system of IoT. malicious cloud services those employed on IoT based network can produce right to use confidential information being transmitted to the preferred target.

10. **Neighbor Discovery:** As discussed above IoT is a collection of devices so IoT designed architecture needs each and every connected device must be identified uniquely. Particularly message communication base authentication has to be secure, undertaking that the data correctly transmit to a device in continuous communication. In neighbor discovery steps like address resolution and router discovery without suitable authentication may capture severe allegations.

## Security Solution for Routing, Transport and Session Security Issues

The threats that take place from reply attacks due to the partition of packets in 6LoWPAN can deal through timestamp and nonce options. These packets are further given to 7LoWPAN version layer subsequent to segment packets. Both timestamp as well as nonce values employed for unidirectional as well as bi-directional packets correspondingly (Kim et al., 2008). (Riaz et al., 2009) projected a framework through secure neighbor discovery, key generation, data encryption, and authentication. Elliptic curve cryptography (ECC) engaged for secure node discovery, the encrypted data make sure node-to-node protection. (Dvir et al., 2011) designed a method for mitigating adversary attacks for the period of map-reading through IPv6 routing protocol for Low-power Lossy networks. The author constructs the directed acyclic graph through root to any gateways. (Weekly et al., 2012) designed a model to cope with sinkhole attacks. On behalf of rank confirmation consequent to Destination Information Object (DIO) message, in one-direction base hash-function employ mutually with a hash series task. In (Hu et al., 2005) authors incorporating symmetric key cryptographic algorithms to make safe nodes starting the attacked wireless nodes in a specific network. Authors in (Wang et al., 2008) describe a procedure to detect wormhole attacks in WSNs by calculating broadcasting distances estimated among neighbors. (Kothmayr et al., 2012) employed a method in favor of end-to-end safety with shared validation in the course of public-key cryptography. Authentication performed via authenticated platform module. Chips using RSA or exit on the trusted Platform module. Here RSA certificates are broadcast in X.509 design. (Park et al., 2016) employed a mutual validation scheme for safe and sound session supervision via asymmetric key-based encryption technique. In the planned system at first pick, an arbitrary number also achieve encryption and create a session key which was consequently utilized for encryption of one more arbitrary integer. The encrypted value is at that moment used for validation. In support of each fresh session, innovative session keys create exclusive of necessitating recurrence of parameters.

## Application Layers Security Issues

IoT based application services helpless against some specific type attacks that are mentioned below.

1. **Insecure Interfaces:** In order to access IoT services mobile, web and cloud interfaces are employed but they vulnerable to dissimilar attacks that cloud rigorously distress the security and privacy of data.
2. **Middleware Security:** Middleware service in IoT planned to deliver message amongst mixed entities of IoT model that must be protected sufficient for the condition of services. Dissimilar interface atmospheres with middleware should be incorporated to endow with secure communication.
3. **CoAP Security with the Internet:** Constrained Application Protocol as web transfer protocol with various security models produces end-to-end safety measures. CoAP follows a special kind of encrypted message format of RFC-7252.

## Security Solution for Application Layers Security Issues

(Liu et al., 2014) planned a scheme for middleware server which maintains data filtering throughout communication along with assorted IoT atmospheres. Projected middleware supports proficient method for naming, addressing and profiling across assorted surroundings. In (OneM2M 2014) authors presents a method for machine-to-machine communications in the IoT atmosphere, standard design with dissimilar layers for security. An energy-efficient security replica using public-key cryptography for IoT based CoAp is offered by (Sethi et al., 2012). It executes through an archetype that utilizes a Mirror proxy (MP) and Resources directory represents server-to-server request during sleep state and the list of assets on the server in that order. (Granjal et al., 2013) proposed another approach of secure message applications communicating through the internet using various CoAp securities. In (Brachmann et al., 2014) IoT based Ip networks, a protection model with 6LBR being used for message filtration in order to afford end-to-end security.

*Table 1. Mapping of IoT layers with security issues/attacks and solutions*

| Layers | Security Issues/Attacks | Solutions |
|---|---|---|
| Physical and Datalink layers | • Sybil attacks<br>• Sleep deprivation attacks<br>• Jamming attacks<br>• Insecure initialization<br>• Insecure physical surface | • Signal strength measurements<br>• Channel estimation<br>• Multi-layer-based IDS<br>• Compute packet delivery ratio<br>• Encode packets with Error-correcting codes |
| Routing, Transport, Session layers | • Session establishment<br>• Wormhole<br>• Sinkhole<br>• Reply attack<br>• Authentication<br>• Buffer reservation attack<br>• Routing attack<br>• Privacy on cloud<br>• Neighbor discovery<br>• End to end security | • Authenticate with ECC, HASH, RSA, SHA1, AES<br>• Apply timestamp, Nonce values<br>• Signature base identification, monitoring node behavior<br>• Graph travels, Random walks on graphs, measuring signal strength |
| Application layer | • Middleware security<br>• CoAP security with Internet<br>• Insecure interfaces | • Key management among devices<br>• Test the interface vulnerabilities with testing tools, firewalls<br>• CoAP mapping, Mirror proxy, Resource directory<br>• Restrict low passwords |

## BLOCKCHAIN SECURITY

Blockchain technology is one of the promising and emergent notions of decentralizes security. It professionally restores a variety of organizations economic transaction systems; moreover, based on mentioned sentences it has the capability of restoring heterogeneous business procedures of various industries. However, it ensures a scattered secured framework to assist exchange, allocation, and integration of information with all abusers and devices, so designers along with decision-makers have to explore it in-depth for its suitability in their respective industry and business appliances (Minhaj et al., 2018; Mahadi et al., 2018).

Blockchain noted as big revolution technology in the present era, in the year of 2008 idea of blockchain was originally initiated by well-known researcher namely Satoshi Nakamoto who executed digital currency famous as Bitcoin. Since decade's people involved in information transmit and transfer of money, property through an online transaction using internet connectivity. These transactions require a trusted third party who has taken responsibility for secured exchange and accountable for any kind of security violation. Coming to the blockchain paradigm, removes central authority among many parties performs fiscal and information transactions with the inclusion of immutable, decentralized public ledger. Here public ledger is a decentralized scattered database which was mutual among the entire network entities. Moreover, it is a cryptographically encrypted, everlasting and tamperproof witness of entire dealings that happen at any time among network entities (Figure 1). Participating entities can able to outlook transaction those associated with them any time they need. Nevertheless, at one time the transaction validated and

*Figure 1. Pictorial representation of blockchain characteristics*

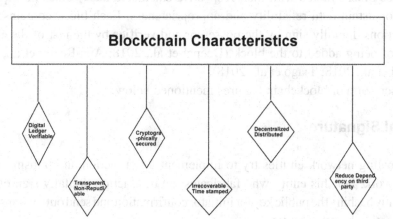

linked with blockchain, it cannot be modified or deleted which turns blockchain undeniable and unalterable. Every transaction confirmed and verified by network entities through pre-defined confirmation or consent mechanisms without any kind of authentication from a central authority. There are numerous advantages with the mentioned sentences, reduces the cost of transmission and abolishes information loss due to single point malfunction because ledger copies distributed as well as synchronized across the network entities. Depending on data management and what actions can be preceded by abusers - blockchain separated into public, private and consortium blockchain. In a public blockchain, which is also acknowledged as a permission-less blockchain, anyone can join without need to take any permissions from third parties. Bitcoin, Litecoin and Ethereum are some instances of a public blockchain. In a private blockchain, a permissioned blockchain, the network owner has made a restriction on network access. In order to be a consortium and not a private chain, the participating companies must be uniformly concerned in the consensus and the decision-making processes of the chain. Technically, consortium chains can be simply executed using solutions such as parity (Tiago et al., 2018; Mahadi et al., 2018).

Consider a simple instance where a group of network entities allocating information and performing exchanges of assets. In place of a trusted third party, network entities make an agreement on a protocol that is to say consensus problem which guarantees for mutual trust, permits validated transaction from peer to peer basis. For the reason, this consensus protocol, digital signatures, proof of work, network entities, cryptographically generated hash values considered as building blocks for Blockchain technology. Here participating network entities could be persons, associations or organizations exchanges a replica of the ledger that contains relevant genuine transactions in chronological order. The public ledger is a collection of sequence blocks linked with their respective calculated hash values in sequential order to sustain data reliability and appropriateness. Each block contains set of transactions digitally sign by the possessor and verifies by the rest of the entities ahead of being added to the block (Deepal et al., 2018; Ana Reyna et al., 2018; Minhaj et al., 2018; Tiago et al., 2018).

Description of blockchain features mentioned below:

## Digital Signature

Participating network entities try to implement a transaction and transmit across the network. For this entity who fabricates the transaction digitally sign on it by constantly hashing the public key for initiator confirmation and send out for transaction confirmation by other nodes.

## Consensus

Consensus plays the main role in blockchain because blockchain entirely works on decentralization, no trustable third hand accountable for the safe transaction, management, and security attacks. New block after collecting transactions starts working on consensus protocol to recognize the verification of the transaction. Proof of work, proof of stake is essential consensus problems.

## Proof of Work and Stake

Proof of work and Stake are confirmation methods for making sure the blockchain transactions. In proof of work, values explored from a pool of vales building the cryptographic hash value of the block, start with the N quantity of zeros in a competition. Coming to proof of stake process, network participants invest digital coins in the blockchain network that represents their stake in a particular block.

## Cryptographic Hash

Proof of work done by a respective network entity, the block transmitted to all the remaining entities who allow it by add to their blockchain after computing a cryptographic hash as Secure Hash Algorithms, for the block Hash utilized as proceeding for the next block.

## Centralized

Blockchain technology completely centralized means there is no inclusion of the third party in any kind of transaction. A public blockchain is decentralized; a private blockchain is fully centralized. Finally, consortium blockchain is moderately centralized.

## Immutability

In public blockchain, transactions stored in dissimilar nodes in a dispersed manner hence is not feasible to tamper. Coming to private and consortium respective governing organizations may possibly tamper the blockchain.

## Efficiency

Security as the major concern frontier of blockchain would be stricter if a large number of nodes takes part in the blockchain. It takes ample time to broadcast transactions and blocks. Hence, the result of transaction throughput is low in addition latency is high. With fewer consensus protocols blockchain could be more adequate.

## Read Permission

All transaction in public blockchain visible publicly whereas read permissions depend on a private blockchain or consortium blockchain.

# CONSENSUS, MINING, MESSAGE VALIDATION MECHANISMS IN BLOCKCHAIN

For accurate execution of Blockchain technology, "Consensus or agreement or Harmony" plays a very important role. The consensus is nothing but a mechanism that decides conditions to be satisfied to signal about an agreement has been accomplishing validations of the blocks to add with blockchain. This will remain you Byzantine Generals Problem, as it needs to make conformity regarding somewhat amongst numerous parties those not having faith on each-other. In crypto-currency, this computational trouble is associated with double-spend trouble, that compact with how to authenticate a little quantity of digital-based currency which was not at all utilized not including proper validation from a trusted third party that regularly maintains a record of all transactions of users. Providing equal consensus mechanism has controlled environment with consists all the miners have the same weights and decide according to the majority of votes. However, it will lead to a Sybil attack where unique abuser with numerous identities can intelligent to organize entire blockchain. In decentralized blockchain architecture, each abuser randomly preferred to insert block, although this random preference may easily vulnerable to attacks. Proof of Work consensus mechanisms proposed to avoid these kinds of attacks, where a node performs heavy work for the network there is a chance for it will not go under any attack. The mentioned work may involve some calculations until a satisfying solution may found. Just the once the mentioned problem is the answer, it is simple for the remaining nodes to validate that gained answer was acceptable. Nevertheless, these type of operations makes blockchain inadequate in energy expenditure, scalability, and throughput which was not applicable in an IoT network where it consists limited resources constrained devices, sensor nodes, etc. (Tiago et al., 2018).

## Proof of Stake

In Proof of work (PoS) based blockchain, entities those have an additional contribution to the network has fewer concerned in any attack on it. Means miners need to verify repeatedly that they have a positive number of participations in the network. In view of the fact that those entities which have continuous participation have control over blockchain. For instance, Peers coin consensus algorithm which takes coinage as a preferred point. Those entities which have oldest, as well as largest groups of coins, would likely to mine a block. Furthermore, Proof of stake consensus problem requires limited computational power compared with proof of work (Peercoin, 2018).

## Proof of Activity

Proof-of-Activity (PoA) consensus algorithm anticipated overcoming the limitations of the problem of Proof-of-stake based on stake age. It builds things even the node is not associated with the set-up; it encourages both ownership and action on the blockchain (Bentov et al., 2014).

## Proof of Stake Velocity (PoSV)

Works based on perception regarding the velocity of money, implemented by Reddcoin. Concepts indicate how many times a unit of currency flows throughout a financial system and group of society members for a certain amount of time. As a rule, the higher the velocity of money the more transactions in which it is used and healthier the economy.

## Transaction as Proof-of-Stake

It is somehow related to PoS. Usually, in PoS Selected nodes contribute for Consensus but in TaPoS every node created transactions supply to the safety of the network.

## Delegated Proof-of-Stake

DPoS is related to PoS, in place of stockholders for selecting and certify blocks; they select certain delegates which can change block size and gaps to do it. By doing fewer nodes involved in block validation, the transaction will be performed faster than other schemes (DPoS 2018).

## PBFT

PBFT consensus algorithms implemented to resolve the Byzantine Generals Problem in the view of asynchronous environments. For adding a new block to the chain, ledger selects the transactions that supported by at least 2/3 rd of all existing nodes (Castro et al., 1999).

## Delegated BFT (DBFT)

It is considered as one of the variants of BFT, performs as a parallel method to DPoS. A number of selected nodes participated in voting to be the ones generate as well as authorize blocks.

## Ripple Consensus Algorithm (RCA)

RCA planned to trim down the high latencies initiated in numerous blockchains which takes place in the employ of synchronous connections between nodes. Ripples server mainly lean on a most authenticated subset nodes while shaping consensus problem (Schwatz et al., 2014).

## Stellar Consensus Protocol (SCP)

SCP is an initial achievement regarding consensus algorithm entitled as Federated Byzantine Agreement (FBA). Performs the same as PBFT but in place of nodes waits or a majority of all other nodes queries, SCP waits only for subset participants which consider being important (Miaziery et al., 2018).

## BFTRaft

It is BFT consensus algorithm planned and designed based on Raft algorithm, trouble-free and straightforward to understand in favor of everyone. BFTRaft combinations improve Raft method by building it towards Byzantine fault-tolerant and by amplifying its security next to diverse threats (Copeland et al., 2018).

## Sieve

IBM based research group implement the Sieve consensus algorithm for hyper ledger-Fabric. The main objective behind the Sieve is run non-deterministic smart cards on permission base scenario. While developing sieve reproduces the procedure associated with non-deterministic smart contracts after that evaluates the outcome. For suppose if variance observed between outcomes achieved with the undersized quantity of processes, those sieved out, if variation excessive entire operation to be sieved out (Cachinc et al., 2016).

## Tendermint

In tender mint consensus algorithm network entities acknowledged as validates moreover recommend blocks of transactions plus make a vote on them. Validation of block done in pre-vote as well as pre-commit. It can merely commit when more than 2/3 validations pre-commit in the round (Kwon et al., 2018).

## Bitcoin-NG

Implemented as a variation of Bitcoin consensus protocol which is mainly designed to improve latency, scalability along with throughput (Eyal et al., 2016).

## Proof of Burn

PoB harmony algorithm entails miners to explain a proof of their loyalty to mining, by means of burning a few crypto-currencies in the course of an unspendable address (Borge et al., 2017).

## Proof of Personhood

PoP harmony/consensus method that constructs utilize ring signature moreover combined symbol to connect substantial to practical identities, in this approach secrecy conserved.

## BLOCKCHAIN AND INTERNET OF THINGS INTEGRATION

The growing development in industrial scientific fields builds a new-fangled method known as the Internet of Things (IoT). IoT is converting and optimizing hand-based-operated processing's which made them as one of the parts in the Digital World where fabricated knowledge base data shows some outcome result. The fabricated data smooth the progression in smart applications enrichment and supervise quality life of citizens all the way in the course of the digitization services in modern cities. From past days cloud commuting technology contribute and produces analyzing results in real-time IoT services with essential functionality toward examine and progression information and finally spin into knowledgeable information (Ana et al., 2018: Diaz et al., 2016) On the other side, in many scenarios vulnerabilities has come to mind because lack of confidence level of systems. Cloud computing used as centralized architecture extensively added to the enlargement of IoT. Nevertheless, coming to data transmission and intelligibility they act as black boxes, more importantly, the applicants of IoT services do not know how and where the data/information which was produced by them is going to be used. Blockchain can able to improve the IoT with the help of faith sharing services, where information is trustworthy and can be observable.

Network participant's data resources talented to be recognized at any time furthermore data remain not able to be forfeited in due course and intensifying its safety measures. In some special scenarios where the IoT information must be steadily shared with numerous participants would symbolize a key riot. Furthermore, sharing consistent data could help the insertion of fresh applicants into the network and contribute to progress their services, acceptances. Hence, the utilization of blockchain can balance IoT through trustworthy and protected information. Many researchers found that IoT can significantly advantage from the functionality afforded by blockchain in addition help advanced future IoT technologies. There are soundless numerous research confront and open problems that to be considered in order to faultlessly utilize these two technologies mutually and this research focal point is still at the initial phase (Clemence et al., 2018; Deepak et al., 2018; Minhaj et al., 2018).

## Blockchain – IoT Integration Models

Before the integration of blockchain with IoT, it is important to mention where transactions occur. Inside the IoT, among IoT - blockchain and hybrid blueprint are some of the integrations involved in IoT, blockchain (Ana et al., 2018; Khan et al., 2017) (See Figure 2).

## IoT - IoT

Generally, IoT devices engage to discover other IoT devices and routing mechanisms. Here some pieces of IoT information accumulate in blockchain, which enables IoT communications to happen exclusive of blockchain technology involvement. This might be the greatest method within words of latency as well as security.

*Figure 2. Symbolic demonstration of blockchain IoT integration models*

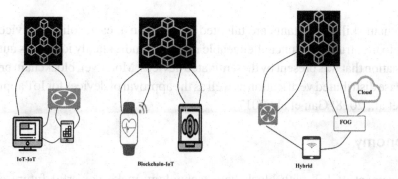

## IoT - Blockchain

Here all transactions move all the way through blockchain, which facilitates an unchangeable record of interactions. It guarantees all selected interactions able to traceable as their particulars reserved in the blockchain. However, organizing each and every transaction in the blockchain would increase bandwidth and data usage.

## Hybrid Method

In hybrid method some transaction operated in blockchain and remaining transaction organized and shared among IoT devices. But the problem occurs when which transaction should take place through blockchain and make a decision in the runtime environment. A faultless architectural combination of both IoT and blockchain influence the profits of blockchain furthermore improves real-time IoT exchanges. At present Fog computing introduction extends limitations of blockchain and IoT.

Here are some of the special improvements provided by the integration of Blockchain and IoT (Clemence et al., 2018; Deepak et al., 2018; Minhaj et al., 2018).

## Decentralized and Scalability

In centralized schemes, few powerful network parts manage processing and storage space regarding information of numerous people. Shifts as central construction to decentralized distributed construction eliminate central points of malfunction and blockages. Decentralization can also improve error tolerance and structure scalability. It would trim down the IoT work pressure as well as progress IoT scalability (Ana et al., 2018; Veena et al., 2015).

## Identity

Blockchain skill contestants are talented to distinguish each solitary device. Data feed into the structure is unchallengeable moreover individually identifies authentic information that was present by disseminated devices. Moreover, blockchain provides trust base distributed verification as well as the approval of devices for IoT appliances (Ana et al., 2018; Gan et al., 2017).

## Autonomy

Improvement of IoT with blockchain technology make powerful future coming application features; it makes possibilities to development of smart autonomous resources, hardware services and capable to interact with each other without involving any servers (Ana et al., 2018; Chain, 2017; Filament, 2017).

## Reliability

The information in IoT unchangeable and distributed over the blockchain. Network participants have complete authority about verifying, assurance the authenticity of the information. Machinery improvements make possible reliability and traceability key aspects of blockchain transport toward IoT (Ana et al., 2018).

## Security

Security plays the main role while data in the transmission process, blockchain secures the transmission of data communications in the form of stored records. While transmitting messages blockchain validate them by smart contracts then only it treated as secured communication between two devices. With the help of blockchain technology present, secure protocols of IoT can be optimized (Ana et al., 2018; Modum, 2017; Filament, 2017).

## Improves Market Services

Blockchain inclusion boosts the IoT eco-system services and market services. It improves IoT inter-relationship and access IoT information toward blockchain (Modum, 2017).

## Secure Operation

With the inclusion of blockchain, not able to forfeited data, code, storage will be securely and steadily operated towards IoT devices (Filament, 2017; Modum, 2017).

## BLOCKCHAIN: IOT INTEGRATION CHALLENGES

Internet of Things integration with Blockchain technology is a tricky and challenging procedure. In view of the fact that the main intention of blockchain design is based on the internet state of affairs through commanding computers and this state of affairs is a faraway technique from IoT authenticity. All connections in blockchain digitally noticed so the devices talented to work. Currency has to be outfitted by this functionality.

## Security

Internet of Things (IoT) is a collection of a heterogeneous network; its applications have to contract with different protection issues at special levels. Moreover, IoT scenario consists of various properties such as mobility of nodes, large scale environment, and wireless communication that affect security. Since numerous security attacks and their respective functions effects on IoT, this shows an indication to make even more complicated security for IoT. After performing several kinds of testing operations researchers found that Blockchain technology as suitable equipment to present and fulfill the much-desired protection improvement in IoT. Reliability of data is the most important challenge while integrating IoT with blockchain technology. Data in chains stays Immutable which is the main functionality of blockchain. For instance, while data reach your destination with corruption, blockchain settles as corrupt. Corrupted data in IoT application may raise more problems apart from malicious attacks. Sometimes devices and nodes not able to work properly because of this corrupted data; this problem cannot questionable until device went under examination tasks. In the view of the above-mentioned scenarios IoT devices as well as nodes comprehensively checked and tested prior to their amalgamation by blockchain technology. Moreover, it is better to incorporate practices on the road to notice machine malfunctions in a little time ahead of attacks take place (Ana et al., 2018; Roman et al., 2018; Lopez et al., 2017; Roman et al., 2013).

## Anonymity and Data Privacy

Anonymity and privacy are the two most important terms in IoT applications. For a request, while IoT device is allied with any person as in e-health state, it is a situation where anonymity desires to be assured. Here device has the capability to hide individuality of person while transmitting individual data. In IoT privacy of data has multiple stages starting from collecting information, enlarges to the communications and finally reaches the application stage. Securing devices which means data which was secure and not easily accessible by other peoples with no proper consent is a challenge. Many types of research utilize numerous cryptographic techniques like IPsec, SSL, TLS, DTLS, symmetric and asymmetric encryption methods but these techniques need a large number of resources which are mostly not available in IoT devices. Applying Blockchain technology identity, anonymity problems will reduce to lighten process. Trust is one more key challenging task while integrating IoT with Blockchain technology (Liu et al., 2015; Wang et al., 2010; Fernandez et al., 2017).

## Legal Issues

Blockchain, particularly in the view of virtual currencies, has made huge controversy regarding legitimacy and legality. Based on necessity and functioning, organize elements foremost the network blockchain approach in the structure of public, private along with consortium blockchains. IoT has well-framed features law and data privacy, protection issues. These laws will become more complicated when IoT added with new technologies like Blockchain. Those expansions of innovative laws and standards can relieve official recognition protection characteristics of devices (Ana et al., 2018).

## Storage and Scalability

Blockchain technology not more appropriate if these two words are not included. IoT devices continuously accumulate a huge amount of data in real-time, this restraint symbolizes an enormous difficulty toward its integration by blockchain. In order to trim-down resources usage, present blockchain technologies implementations processed through a few transactions per second. This will potentially blockage for IoT usage with blockchain. Moreover, blockchain doesn't have the power to store data like IoT. Entire data which was accumulated in IoT is not so useful; at present some feature extracting or selecting methods like filters, compression methods which perform great actions. Data compression could reduce transmission, storage space and processing tasks regarding high volume IoT data. Here some focus also needs on blockchain consensus protocols which cause a bottleneck to improve bandwidth and reduce the latency of its communication thereby permit batterer transactions (Ana et al., 2018).

## Smart Contracts

IoT has some benefits on or after employ of smart contracts which depend on approaches where contracts implemented at various levels. Theoretically, the contract is a set of code assignments with data positions that exist in definite blockchain tackles. Smart contracts offer secure along with a reliable system for IoT, by recording and managing all transactions. Smart contracts must take attention on heterogeneous and restraints nearby in the IoT. Filter and cluster methods could be harmonized via smart contracts toward facilitating appliances to deal with IoT depending on circumstance and requirements (Ana et al., 2018).

## Consensus or Harmony

Blockchain has a wide variety of consensus algorithms those immutable and not thoroughly tested as much as necessary. By considering IoT applications restricted resource constraints of nodes as well as devices make them not eligible for taking part in consensus protocols. Mining is another challenge faced by IoT due to its limited resource constraints.

## APPLICATIONS OF BLOCKCHAIN-IOT

Blockchain technology applicable to many fields, applicability starts from Bitcoin then after moving towards smart cards finally progress towards justice, efficiency applications. Bitcoin which is known as Blockchain 1.0 or Digital currency. It indicates mining, hashing, and public ledger, transaction assured software and digital currency that symbolizes a store, protocol value. Introduction of bitcoin greatly reduces the third-party online transaction charges which protect against price increases, bitcoin produce superior security than credit-cards. Blockchains smartcards implementation is also known as Blockchain 2.0 or Digital Economy that symbolizes financial and economic applications like payments of loans in traditional banking systems and transfers of bonds, stocks, derivatives, tiles, contracts, and assets in financial marketing. Smartcard contracts are essentially computer-based programs that can as a result of design complete the terms and conditions of an agreement. At the time of pre-arranged circumstance in a smart contract between two networks entities assemble then the contractual understanding can be mechanically prepared their payments as per the contract in transparent approach. Ethereum is one of the examples for Digital economy. Finally, Blockchain 3.0 or Digital society has numerous applications include science, public goods, identity, art, governance, health and various aspects of culture and communication. (See Figure 3)

## Financial Services

Implementation of Blockchain technology such as Bitcoin made tremendous changes in traditional financial as well as business services. It also makes attention to software companies and legal organizations. Peters et al have mentioned that blockchain has the capability to interrupt world banking working process like as settlement of financial assets etc. Risk management agenda has a major part in financial technology; it performs better with a combination of blockchain technology. Pilkington revealed the

*Figure 3. Symbolic demonstration on Blockchain IoT applications*

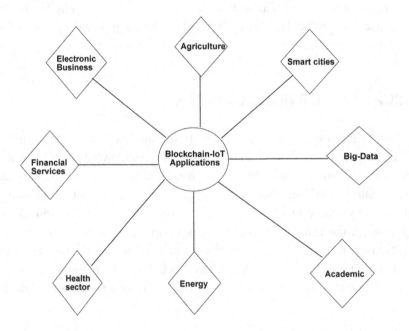

risk management agenda with blockchain that utilized to analyze investment risks. (Noyes 2016) build peer tot peer secure financial market by applying blockchain technology (Tiago et al., 2018; Zibin et al., 2018; Noyes, 2016).

## Agriculture

IoT based agriculture appliances also engaged in Blockchain technology. For instance, many authors present traceability scheme to track food supplies, which is prepared in support of Radio frequency identification and blockchain technology (Tian et al., 2016; Tiago et al., 2018).

## Energy Sector

The energy sector is also one of the applications of Blockchain to IoT. Authors (Lundqvist et al., 2017) design various methods based on blockchain with internet of energy where devices have to pay for services without any human interventions. In proposed methods smart chains that unite with a well-dressed socket which is proficient to pay for the energy consumed.

## Healthcare

In (Bocek et al., 2017) authors designed a traceability application that applies blockchain technology to validate information truthfulness, user-friendliness to high-temperature records in a pharmaceutical stock chain. In (Hardjoin et al., 2016) authors developed the construction of blockchain base proposal for experimental trials along with the accuracy of medicine.

## Big Data

Big data can successes with blockchain technology, researchers developing blockchain-based solutions that can handle collect and control a huge amount of data composed as of IoT networks.

## Smart Cities

In (Biswas et al., 2016) authors presented blockchain-based IoT framework related to smart cities, in (Ahram et al., 2017) various blockchain connections with industrial appliances mentioned.

## Electronic-Business

(Zhang 2015) designed IoT e-business model base on the blockchain, smart contract. Authors developed distributed autonomous corporations (DAC) like centralized operation unit which obtain as well as switch sensor data exclusive of any third party.

## Academics

Academic reputation is important for a better society. (Sharples, 2015) designed a blockchain base scattered scheme for academic records as well as reputation. On the preliminary stage, every institution, as well as workers, provided with an opening reward of educational reputation currency. Institution/Organization can reward or transfer a few standing records to the staff, all records accumulate in blockchain so any kind of reputation changes can easily be detected.

## Energy Saving

(Gogerty et al., 2011) designed blockchain-based green energy that encourages handling of renewable sources. Solarcoin distributed by sloarcoin foundation as long as solar energy generated by you, here solar coin is one type of digital currency incentive for solar power creators.

## Safety, Privacy

Both safety and privacy defense important concern in IoT services. Inclusion of blockchain improves security and privacy in IoT based appliances. (Hardjoin et al., 2016) designed privacy-preserving method for IoT devices in a cloud environment. Authors also developed a method where devices to improve manufacturing attribution without any authentication from third-party.

## FUTURE CHALLENGES, DIRECTIONS, AND ISSUES

Blockchain and Internet of Things exposed their prospective in production, commerce and academic (Tiago et al., 2018; Minhaj et al., 2018). Some of the future works based on existing literature survey are mentioned below.

## Blockchain Construction

It is necessary to construct a complete trust base framework or infrastructure which able to carry-out every necessity to making use of blockchain in IoT services. For this reason, researchers need to develop numerous high rated methods which control inter-domain issues.

## Complicated Technical Challenges

Both IoT and Blockchain reach their advanced levels but technically there still some issues to be solved such as privacy, security, scalability, cryptographic techniques development and stability requirements in BIoT appliances. Furthermore, blockchain technology various design, limitation issues in consensus protocols, implementations in smart cards and transaction capacities.

## Rapid Testing

In the future when users may wish to merge blockchain and IoT systems. For this merger process, the initial pace will be make out which kind of blockchain fits for the user requirements; hence it is compulsory to create a system to test dissimilar blockchains.

## Regulatory and Legal Aspects

Determining regulatory standards such as ownership, national and international jurisdiction are the principal problems to be solved beside technical challenges that can expand the promising value of BIoT.

## Hardware Vulnerabilities

IoT is more vulnerable to hardware attacks because of low-cost and low-power devices. Routing, as well as packet preprocessing mechanisms, must be verified before employed in IoT. It is a tricky operation to detect vulnerabilities after network deployment so there is need of standard verification protocol which can exploit the IoT security.

## Blockchain Vulnerabilities

Even though advanced methods proposed for IoT security, there is a chance for blockchain to be vulnerable. The consensus protocols depending on miners hashing rule can be concessional which allows the invader to maintain the entire blockchain. Effective mechanisms still are distinct to make sure the privacy of transaction dealings moreover stay away from attacks that might produce double-spending for the duration of transactions.

## Interoperability of Security Protocols

In order to provide universal defense mechanisms for IoT, the designed protocols at special layers interoperate with each other for translation mechanisms. Inside that universal defense mechanism, each layers valuable arrangement of defense principles specific throughout consideration of architectural limitations.

## Resource Limitations

IoT has a resource-constrained architecture which is the main issues to define a robust security mechanism. Conventional cryptographic procedures work with these limited resource constraints, moreover interchange of keys certificates as well as storage purpose more energy requirement is needed. Hence unbeaten execution of security, as well as message communication protocols in favor of IoT, needed which works efficiently in the limited energy-constrained devices, nodes.

## Trusted Management and Software Update

IoT works with an outsized amount of devices for providing scalable and trusted management with a secure software update in outsized amount IoT devices is still one of the open research problems. Blockchain çan ensure some of those security issues; on other hand blockchain technology suffers from scalability, key collision, and efficiency.

## CONCLUSION

The purpose of the Blockchain concept has enlarged ahead of it's utilizing for Bitcoin production as well as transaction dealings. At present Blockchain along with its modifications are utilized to protect any type of transactions, rather It Human-

To-Human or Machine-To-Machine connections. Blockchain adoption appears to be revolutionizing the IoT. This chapter inspects the up-to-date of blockchain-IoT equipment and projected important circumstances for BIoT appliances in areas like big-data, healthcare, financial services, agriculture, and energy management. Moreover, BIoT integration issues, challenges, applications, mining, and message validation methods explained. However, BIoT state of affairs facade specific technical requirements that vary from execution such as energy effectiveness in resource-constrained procedures or the requirement of precise structural design.

## REFERENCES

Bentov, I., Lee, C., Mizrahi, A., & Rosenfeld, M. (2014, June). Proof of activity: Extending bitcoin's proof of work via proof of stake. In *Proceedings of the 9th Workshop on the Economics of Networks, Systems and Computation*, Austin, TX.

Blockchain of things. (2017). Available online: https://www.blockchainofthings. com (*Accessed 1 February 2018*)

Borge, M., Kokoris-Kogias, E., Jovanovic, P., Gasser, L., Gailly, N., & Ford, B. (2017, April). Proof-of-personhood: Redemocratizing permissionless cryptocurrencies. In *Proceedings of the IEEE European Symposium on Security and Privacy Workshop*, Paris, France.

Brachmann, M., Garcia-Morchon, O., Keoh, S.-L., & Kumar, S. S. (2012). Security considerations around end-to-end security in the IP-based internet of things. In *Proceedings 2012 Workshop on Smart Object Security, in Conjunction with IETF83*, pp. 1–3.

Cachin, C., Schubert, S., & Vukolic, M. (2016, December). Non-determinism in Byzantine´ fault-tolerant replication. In *Proceedings of the International Conference on Principles of Distributed Systems (OPODIS 2016)*, Madrid, Spain.

Castro, M. & Liskov, B. (1999, February). Practical Byzantine fault tolerance. In *Proceedings of the Third Symposium on Operating Systems Design and Implementation*, New Orleans, LA.

Conzon, D., Bolognesi, T., Brizzi, P., Lotito, A., Tomasi, R., & Spirito, M. A. (2012). The VIRTUS middleware: An XMPP based architecture for secure IoT communications. In *Proceedings 2012 21st International Conference on Computer Communications and Networks, ICCCN*. doi:.6289309. pp 1-6.10.1109/ICCCN.2012

Copeland, C. & Zhong, H. (2018). *Tangaroa: A Byzantine fault tolerant raft*. Available at http://www.scs.stanford.edu/14au- cs244b/labs/projects/copeland_zhong.pdf

Demirbas, M. & Song, Y. (2006). An RSSI-based scheme for sybil attack detection in wireless sensor networks. In *Proceedings of the 2006 International Symposium on World of Wireless, Mobile and Multimedia Networks, WOWMOM '06*, IEEE Computer Society, Washington, DC, pp. 564–570. 10.1109/WOWMOM.2006.27

M. Díaz, C. Martín, B. Rubio. (2016). State-of-the-art, challenges, and open issues in the integration of internet of things and cloud computing. *Journal of Networks Computers*. pp. 99-117.

Dvir, A., Holczer, T., & Buttyan, L. (2011). VeRA - version number and rank authentication in RPL. In *2011 IEEE Eighth International Conference on Mobile Ad-Hoc and Sensor Systems*. pp. 709-714.

Eyal, I., Gencer, A. E., Sirer, E. G., & Van Renesse, R. (2016). Bitcoin-NG: a scalable blockchain protocol. In *Proceedings 13th USENIX Symposium on Networked Systems Design and Implementation (NSDI 16)*, Santa Clara, CA, pp. 45–59.

Eyal, I., Gencer, A. E., Sirer, E. G., & Van Renesse, R. (2016, March). Bitcoin-NG: A scalable blockchain protocol. In *Proceedings of the 13th USENIX Symposium on Networked Systems Design and Implementation*, Santa Clara, CA.

Fernandez-Gago, C., Moyano, F., & Lopez, J. (2017). Modelling trust dynamics in the internet of things. *Information Science*, *396*, 72–82. doi:10.1016/j.ins.2017.02.039

Filament. (2017). Available at https://filament.com/

Gan, S. (2017). *An IoT Simulator in NS3 and a Key-Based Authentication Architecture for IoT Devices using Blockchain*. Indian Institute of Technology Kanpur.

Gmez-Goiri, A., Ordua, P., Diego, J., & de Ipiña, D. L. (2014). *Otsopack: Lightweight semantic framework for interoperable ambient intelligence applications* (pp. 460–467). Computer Human Behaviour. doi:10.1016/j.chb

Granjal, J., Monteiro, E., & Silva, J. S. (2013). Application-layer security for the WoT: Extending CoAP to support end-to-end message security for internet-integrated sensing applications. In *Proceedings International Conference on Wired/Wireless Internet Communication, Springer Berlin Heidelberg*, pp. 140–153. 10.1007/978-3-642-38401-1_11

Hu, Y.-C., Perrig, A., & Johnson, D. B. (2005). Ariadne: A secure on-demand routing protocol for ad hoc networks. *Wireless Networks*, *11*(1), 21–38. doi:10.100711276-004-4744-y

Khan, M. A. & Salah, K. (2018). IoT security: Review, blockchain solutions, and open challenges. *Future Generation Computer Systems*. pp. 395–411.

Khan, M. A., & Salah, K. (2017). *IoT security: Review, blockchain solutions, and open challenges. Future Generation Computer Systems*.

Kim, H. Protection against packet fragmentation attacks at 6LoWPAN adaptation layer. In *Proceedings 2008 International Conference on Convergence and Hybrid Information Technology.* . pp. 790-801.10.1109/ICHIT.2008.261

Kothmayr, T., Schmitt, C., Hu, W., Brnig, M., & Carle, G. (2012). *37th annual IEEE Conference on Local Computer Networks - Workshops.* doi:. pp. 964–972.10.1109/LCNW.2012.6424088

Kwon, J. (2018). Tendermint: Consensus without mining (v0.6). Available at https://tendermint.com/static/docs/tendermint.pdf

Larimer, D. (2018). *Transactions as proof-of-stake.* Available at https://bravenewcoin.com/assets/Uploads/TransactionsAsProofOfStake10.pdf

Li, Q., & Trappe, W. (2006). Light-weight detection of spoofing attacks in wireless networks. In Proceedings *2006 IEEE International Conference on Mobile Ad Hoc and Sensor Systems*, pp. 845–851. 10.1109/MOBHOC.2006.278663

Liu, C., Ranjan, R., Yang, C., Zhang, X., Wang, L., & Chen, J. (2015). Mur-dpa: Top-down levelled multi-replica merkle hash tree based secure public auditing for dynamic big data storage on cloud. *IEEE Transactions on Computers*, *64*(9), 2609–2622. doi:10.1109/TC.2014.2375190

Liu, C., Yang, C., Zhang, X., & Chen, J. (2015). External integrity verification for outsourced big data in cloud and IoT: A big picture. *Future Generation Computer Systems*, *49*, 58–67. doi:10.1016/j.future.2014.08.007

Liu, C. H., Yang, B., & Liu, T. (2014). Efficient naming, addressing and profile services in Internet-of-Things sensory environments. *Ad Hoc Networks*, *18*, 85–101. doi:10.1016/j.adhoc.2013.02.008

Lopez, J., Rios, R., Bao, F., & Wang, G. (2017). Evolving privacy: From sensors to the internet of things. *Future Generation Computer Systems*, *75*, 46–57. doi:10.1016/j.future.2017.04.045

Lundqvist, T. & De Blanche, A. (2017). Thing-to-thing electricity micro payments using blockchain technology. In *Proceedings of the Global Internet of Things Summit (GIoTS),* Geneva, Switzerland, pp. 6-9.

Mazieres, D. (2018). *The stellar consensus protocol: A federated model for internet-level consensus.* Available at papers/stellar-consensus-protocol.pdf

Modum. (2017). Available at https://modum.io/

Noubir, G., & Lin, G. (2003). Low-power DoS attacks in data wireless LANs and countermeasures, SIGMOBILE Mob. *Computer Communication Review*, *7*(3), 29–30. doi:10.1145/961268.961277

Noyes, C. (2016a). Bitav: Fast anti-malware by distributed blockchain consensus and feedforward scanning. *arXiv preprint arXiv:1601.01405*

OneM2M. (2017). Security solutions–OneM2M technical specification. Retrieved from http://onem2m.org/technical/latest-drafts

Ongaro, D. & Ousterhout, J. (2014, June). In search of an understandable consensus algorithm. In *Proceedings of USENIX Annual Technical Conference,* Philadelphia, PA.

Park, N., & Kang, N. (2016). Mutual authentication scheme in secure internet of things technology for comfortable lifestyle. *Sensors (Basel)*, *6*(1), 20–20. doi:10.339016010020 PMID:26712759

Pecorella, T., Brilli, L., & Muchhi, L. (2016). The role of physical layer security in IoT: A novel perspective. *Information, 7*(3).

Peercoin official web page. Available at https://peercoin.net

Prisco, G. (2016) Slock. it to introduce smart locks linked to smart ethereum contracts, decentralize the sharing economy. Bitcoin Magazine. Nov-2015 [Online]. Available at https://bitcoinmagazine.com/articles/sloc-it-to-introduce-smart-locs-lined-to-smart-ethereum-contractsdecentralize-the-sharing-economy-1446746719.

Puthal, D., Malik, N., Mohanty, S. P., Kougianos, E., & Yang, C. (2018). Blockchain as decentralized security framework. *IEEE Consumer Electronics Magazine* March 2018. DPOS description on Bitshares. Available at http://docs.bitshares.org/ bitshares/ dpos.html

Ren, L. (2018). Proof of stake velocity: Building the social currency of the digital age. Available at https://www.reddcoin.com/papers/PoSV. pdf

Reyna, A., Martín, C., Chen, J., Soler, E., & Díaz, M. (2018). On blockchain and its integration with IoT, Challenges and opportunities, *Future Generation Computer Systems*, 88, pp. 173-190.

Riaz, R., Kim, K.-H., & Ahmed, H. F. (2009). Security analysis survey and framework design for IP connected LoWPANs. In *Proceedings 2009 International Symposium on Autonomous Decentralized Systems*, pp. 1–6. 10.1109/ISADS.2009.5207373

Roman, R., Lopez, J., & Mambo, M. (2018). A survey and analysis of security threats and challenges. *Future Generation Computer Systems*, *78*, 680–698. doi:10.1016/j.future.2016.11.009

Roman, R., Zhou, J., & Lopez, J. (2013). On the features and challenges of security and privacy in distributed internet of things. *Computer Networks*, *57*(10), 2266–2279. doi:10.1016/j.comnet.2012.12.018

Schwartz, D., Youngs, N., & Britto, A. (2014). The ripple protocol consensus algorithm. *White paper, Ripple Labs*.

Sethi, M., Arkko, J., & Kernen, A. (2012). End-to-end security for sleepy smart object networks. In *Proceedings 37th Annual IEEE Conference on Local Computer Networks Workshops*, pp. 964–972. 10.1109/LCNW.2012.6424089

Tiago, M. F.-C. & Fraga-Lamas, P. (2018). A review on the use of blockchain for the Internet of Things. *IEEE Transactions*.

Tian, F. (2016, June). An agri-food supply chain traceability system for china based on RFID & blockchain technology. In *Proceedings of the 13th International Conference on Service Systems and Services Management*, Kunming, China.

Veena, P., Panikkar, S., Nair, S., & Brody, P. (2015). Empowering the edge-practical insights on a decentralized internet of things. *IBM Institute for Business*, 17.

Wang, C., Wang, Q., Ren, K., & Lou, W. (2010). Privacy-preserving public auditing for data storage security in cloud computing. In *Proceedings INFOCOM, San Diego*, CA, IEEE, 2010, pp. 1–9. 10.1109/INFCOM.2010.5462173

Wang, W., Kong, J., Bhargava, B., & Gerla, M. (2008). Visualisation of wormholes in underwater sensor networks: A distributed approach. *Int. J. Secur. Netw.*, *3*(1), 10–23. doi:10.1504/IJSN.2008.016198

Weekly, K. & Pister, K. (2012). Evaluating sinkhole defense techniques in RPL networks. In *Proceedings of the 2012 20th IEEE International Conference on Network Protocols (ICNP)*, IEEE Computer Society, Washington, DC. . pp 1-6. 10.1109/ICNP.2012.6459948

Xiao, L., Greenstein, L. J., Mandayam, N. B., & Trappe, W. (2009). Channel-based detection of sybil attacks in wireless networks. *IEEE Transactions on Information Forensics and Security*, *4*(3), 492–503. doi:10.1109/TIFS.2009.2026454

Xu, W., Trappe, W., Zhang, Y., & Wood, T. (2005). The feasibility of launching and detecting jamming attacks in wireless networks. In *Proceedings of the 6th ACM International Symposium on Mobile Ad Hoc Networking and Computing, nMobiHoc '05*, ACM, New York, NY. . pp. 46–57.10.1145/1062689.1062697

Young, M., & Boutaba, R. (2011). Overcoming adversaries in sensor networks: A survey of theoretical models and algorithmic approaches for tolerating malicious interference. *IEEE Communications Surveys and Tutorials*, *13*(4), 617–641. doi:10.1109/SURV.2011.041311.00156

Zheng, Z., Xie, S., Dai, H.-N., Chen, X., & Wang, H. (2018). Blockchain challenges and opportunities: A survey. *International Journal of Web and Grid Services*, *14*(4), 352. doi:10.1504/IJWGS.2018.095647

# Chapter 11
# Secure System Model for Replicated DRTDBS

**Pratik Shrivastava**
*Madan Mohan Malaviya University of Technology, India*

**Udai Shanker**
*Madan Mohan Malaviya University of Technology, India*

## ABSTRACT

*Security in replicated distributed real time database system (RDRTDBS) is still explorative and, despite an increase in real-time applications, many issues and challenges remain in designing a more secure system model. However, very little research has been reported for maintaining security, timeliness, and mutual consistency. This chapter proposes the secure system model for RDRTDBS which secures the system from malicious attack. To prevent the request/response from malicious attack, authors have extended the system model with a cryptographic algorithm. In the cryptographic algorithm, a key must be secretly known only to the sender and receiver. Thus, in this chapter, authors have used the key generation algorithm to generate a key using an image. This secure system model maintains the confidentiality of the replicated data item and preserves its data integrity. It performs better in terms of malicious attack compared to other non-secure system models.*

DOI: 10.4018/978-1-7998-0373-7.ch011

# INTRODUCTION

The confluence of real time systems, communication network, and database systems is creating a real time database system (RTDBSs). Real time application processing in such an RTDBSs requires timely completion of real time transaction (RTT) such that temporal consistency of real time data item can be maintained. Failing to meet such a demand of real time data item cause heavy economic loss. Thus, the primary focus of RTDBSs is timeliness irrespective of logical consistency. Recently, the demand of RTDBS is expanding rapidly, a large number of real time applications such as stock management system, banking system, business information system, and air traffic control system generates a massive amount of data. A distributed real time database system (DRTDBS) has been specifically designed to satisfy the timeliness demand of RTT such that temporal consistency of such huge amount of data can be maintained. However, due to distributed processing of RTT and following strict consistency criteria has resulted in an increased research effort in this area. The research area includes concurrency control protocol (CCP), commit protocol (CP), replication technique (RT) and buffer management (BM) to maximize majority of RTTs to get successfully complete within their deadline. RT in DRTDBS is usually used to increase the performance of the system in terms of scalability, reliability, availability, and fault-tolerance.

In RDRTDBS, the main concern is to maintain the mutual consistency between data replicas despite a malicious attack from unauthorized users and compromised replicas (Zhao, 2014). This is accomplished via totally ordering the request deterministically, and simultaneously propagating this request to all the master sites or slave sites (Zhang, 2011). Replication protocol (RPL) (i.e. replica concurrency protocol or replica control technique) is used to maintain the mutual consistency between data replica present in such a master sites or slave sites. Existing research has been conducted mainly on designing an effective and efficient RPL (Gustavsson, et al., 2004; Gustavsson, et al., 2005; Haj, et al., 2008; Kim, 1996; Mathiason et al., 2007; Peddi & DiPippo., 2002; Salem et al., 2016; Shrivastava & Shanker, 2018; Shrivastava & Shanker, 2018; Shrivastava & Shanker, 2019; Shrivastava & Shanker, 2018; Son & Kouloumbis, 1993; Son & Zhang, 1995; Son et al., 1996; Syberfeldt, 2007; Xiong et al., 2002). Additionally, these RPLs were following different correctness criteria such that one copy serializability (1SR) or weaker than 1SR can be satisfied (Ouzzani et al., 2009).

Although existing work (Gustavsson et al., 2004; Gustavsson et al. 2005; Haj et al., 2008; Kim, 1996; Mathiason et al., 2007; Peddi & DiPippo, 2002; Salem et al., 2016; Shrivastava & Shanker, 2018; Shrivastava & Shanker, 2018; Shrivastava

& Shanker, 2019; Shrivastava & Shanker, 2018; Son & Kouloumbis, 1993; Son & Zhang, 1995; Son et al., 1996; Syberfeldt, 2007; Xiong et al., 2002). is acceptable to work for some real-time applications where guaranteeing security is not an important factor but when confidentiality and integrity of replicated data item become important than such existing work will be of no use. For example, consider a real time application that relies on the use of the consistent value of replicated data item. This consistent value is used to take a judgmental decision for organization benefit. However, if these replicated data items get hijacked by an unauthorized user, then would cause disclosure of the valuable information. This may ultimately cause a heavy economic loss in the system. Thus, the confidentially and data integrity of the replicated data item in such a securely demand real time application is necessary.

In the recently published article (Shrivastava & Shanker, 2018), we have identified that existing research in RDRTDBS suffers from the issues of QoS, Security, Dependency Relationship and strict correctness criteria (Ouzzani et al., 2009). Thus, in the current chapter, we concentrate on the security aspect of RDRTDBS which secures the system from malicious attack. RDRTDBS security issues may stem due to the technology implementation, service offering, and from its characteristics. All the participating sites holding replicas must be secure, for an RDRTDBS to be secured. Thus, in RDRTDBS, the security of the system does not solely depend on the individual's replica security measures. The neighboring replica may provide an opportunity to an unauthorized user to access the original data. In RDRTDBS, user access is spread over the distinct location that may result in various security concerns. Unauthorized data access by the unauthorized user must be prevented. However, a key factor determining the performance of an RDRTDBS that stores data replica is the consistent value. Thus, we can deduce that both security and mutual consistency are critical for the next generation small, medium, and large-scale systems.

Existing system model (Gustavsson et al., 2004; Gustavsson et al. 2005; Haj et al., 2008; Kim, 1996; Mathiason et al., 2007; Peddi & DiPippo, 2002; Salem et al., 2016; Shrivastava & Shanker, 2018; Shrivastava & Shanker, 2018; Shrivastava & Shanker, 2019; Shrivastava & Shanker, 2018; Son & Kouloumbis, 1993; Son & Zhang, 1995; Son et al., 1996; Syberfeldt, 2007; Xiong et al., 2002) of RDRTDBS has been mainly designed to maintain the replica consistency, and these RPLs were operating from kernel, thus they are termed as kernel-based RPLs. These system models do not enforce any security algorithm that provides the opportunity to an unintended user to get easy access. To secure these system models from a malicious attack, we can approach two ways: (1) existing system models get extend with security algorithms, or (2) propose a new secure system model. In this chapter, we have extended the system model (Shrivastava & Shanker, 2018) with security techniques which ultimately prevents the unintended user form accessing and modifying the valuable data. The reason for selecting this system model because here processing of maintaining mutual consistency get shifted from kernel to the middleware server. Thus, master/slave sites are free to process only RTT and middleware server will conduct the task of maintaining mutual consistency. In this extended system model, we enforce the cryptographic algorithm (Nie & Zhang, 2009). such that unintended user is unable to get access to the original data and unable to modify its value. To provide such security, we encrypt the admitted real time transaction (RTT) from the client via encryption algorithm (Nie & Zhang, 2009). This Encryption algorithm uses the key which is generated from the key generation algorithm proposed in (Shrivastava et al., 2014). This key generation algorithm takes the input of image and this image is designed according to user creativity. Due to using user designed image, the task of identifying the key from the image by the hijacker becomes more difficult. The reason for such difficulty is because the image contains a large number of pixels value and key generation uses such pixel value to generate the key. Additionally, the client can use the same image or different image for encryption of next RTT, which makes the task of cracking the key from the image more critical. After encryption of RTT in the client site, this encrypted text is forwarded to the middleware, the middleware will decrypt this encrypted message using the key generated from the same image designed by the user. After decryption of encrypted text, middleware processes the request as already defined in (Shrivastava & Shanker, 2018). After processing in the middleware, the middleware will encrypt the RTT using same image designed by the user. If the encrypted RTT is write RTT or update RTT, then middleware will broadcast it to all the master sites. All master sites receive such broadcasted encrypted RTT and decrypt it to process the RTT such that all master sites get synchronously reach to the consistent state. Apart from write RTT or update RTT, if the RTT is read RTT, then middleware will unicast it to any randomly selected master site. The master site will further unicast this encrypted RTT to the least loaded slave site without interpreting the content of encrypted RTT. When the processing of RTT in master/slave site gets completed, the master/slave will encrypt the response and propagate to the client. The client will receive such a message and decrypt it to get the original response. During this whole process, the message is transmitted in the encrypted format in the network such that unintended user is unable to bypass the confidentiality and integrity of the original content.

The rest of this entry presents an introduction where the main concepts of mutual consistency and security are described. Furthermore, the mutual consistency management provided by existing RPLs is reviewed, pointing out their approach to replica synchronization. The core problem of malicious attack in existing system model is presented in the main focus section, where the main issue of maintaining mutual consistency and security is discussed. Extended system model with security algorithms is presented in the solution and recommendation section, and the discussion of our extended system model is presented in the discussion section.

## BACKGROUND

The DRTDBS is a timely constraint database system whom correctness depends on the timely completion of RTTs (Shanker et al., 2008). Missing the deadline of RTT in such a system causes heavy loss. for instance, missing the deadline of hard RTT cause serious harm, missing the deadline of firm RTT leaves no value for the system (Wang et al., 2011), and missing the deadline of soft RTT leaves some value for the system. This tight requirement of satisfying deadline and consistency makes the system more complex. Thus, this system is usually equipped with an RT (Shrivastava & Shanker, 2018, August) to satisfy the timely demand of RTT more linearly. This extended version of DRTDBS with replication technique is termed as RDRTDBS.

In RDRTDBS, data replica is created in distinct sites which increases the opportunity of RTT to get processed locally. Data replicas are non-replicated, partially replicated or fully replicated. The decision for replicas creation will depend upon the RTT type. If the system is read intensive, then the system gets fully replicated. However, if the system is update intensive, then the system gets partially replicated. In fully replicated RDRTDBS, RPL has to maintain the mutual consistency of all the replicas whereas in partially replicated DRTDBS RPL have to maintain the mutual consistency between replica sites that are holding the same data copy. Consequently, both the approaches of replicating the data have their own pros and cons. ViFuR (Mathiason et al., 2007) has solved the issues of both the replicating approaches more effectively. In ViFuR, replicas are virtually fully replicated to all sites such that on-demand required replica is updated to satisfy the data requirement of admitted RTT locally. Although replication technique increases the performance of the system in terms of availability, reliability, fault-tolerance, and scalability. Every such feature depends on the consistency of the replicated data item. In order to maintain the consistency between replicated data object, existing research (Gustavsson et al., 2004; Gustavsson et al., 2005; Haj et al., 2008; Kim, 1996; Mathiason et al., 2007; Peddi & DiPippo, 2002; Salem et al., 2016; Shrivastava

& Shanker, 2018; Shrivastava & Shanker, 2018; Shrivastava & Shanker, 2019; Shrivastava & Shanker, 2018; Son & Kouloumbis, 1993; Son & Zhang, 1995; Son et al., 1996; Syberfeldt, 2007; Xiong, 2002; Gupta & Swaroop, 2018; Lokhande & Dhainje, 2019) has been conducted in the development of efficient and effective RPLs. These RPLs operates on their assumed system models which are either fully replicated, partially replicated or ViFuR. The effectiveness of these RPLs depends on the system model. System model includes master sites, client, and communication network. These constituents affect the performance of RDRTDBS. Thus, we can deduce that in RDRTDBS, both system model and RPLs plays an important role to improve the system performance.

Apart from mutual consistency in RDRTDBS, Security is related to enforce the policy in the system such that respected user can utilize (i.e. read, write, modify, delete) the data as per its authorization. Basically, we have gone through around two policies (i.e. multilevel security policy and discretionary security policy) that can be incorporated to prevent the security issues in DRTDBS. Multilevel security is the integration of clearance level and classification level where clearance level can be in the form of public or secret and classification level can be in the form of top secret, secret, confidential, or unclassified. Despite of multilevel security policy, discretionary security policy prevents access based on the type of access and identity of user. However, such policy may not be capable to prevent unauthorized disclosure of information. Thus, research in the direction of enforcing a cryptography algorithm is necessary that prevents unauthorized access and secures the system.

Overall from this discussion, we can deduce that in RDRTDBS both system model and RPL have a strong relationship to maintain the mutual consistency. Additionally, further research is more required in the direction of enforcing a cryptographic algorithm such that the system can be made secure from malicious attack and, simultaneously mutual consistency, timeliness can be satisfied.

## MAIN FOCUS OF THE CHAPTER

As already mentioned in our previous section that in existing RDRTDBS, the majority of research work in my best of knowledge has been done on maintaining mutual consistency between replicated data. RTT processing on a mutual consistent data generates a consistent result which is important where simultaneously consistency and timeliness are important. Existing research work done for RDRTDBS is acceptable to work for an existing scenario where security is a least an important factor. However, such work is not acceptable to work for future coming real time application where security is an important factor.

Let us consider a scenario of a bank system where RDRTDBS is employed to satisfy the increasing demand from their employee and customers. In this system, suppose a customer admits the request for withdrawing an amount from his account. After processing such request in the system, the output gets commit. To maintain the replica consistency, system propagate the updated log to another replica site to maintain the mutual consistency. If this log record is not encrypted, then such log record may be easily trapped by a hijacker. After trapping, hijacker can easily bypass the security constraint in the system and causes the harm in two ways: (1) hijacker can access the confidential information and sale it another competitor party which may cause heavy economic loss, (2) hijacker can change the log record and propagate this updated log to other replica sites which ultimately bypass the data integrity constraints. Thus, to secure the system from such losses, it becomes necessary to encrypt the log record before transmitting in the network. To carry over this work, we enforce a cryptographic algorithm in the RDRTDBS which maintains the confidentiality and data integrity of the valuable information.

To accomplish this work, we have explored different existing system model (Gustavsson et al., 2004; Gustavsson et al. 2005; Haj et al., 2008; Kim, 1996; Mathiason et al., 2007; Peddi & DiPippo, 2002; Salem et al., 2016; Shrivastava & Shanker, 2018; Shrivastava & Shanker, 2018; Shrivastava & Shanker, 2019; Shrivastava & Shanker, 2018; Son & Kouloumbis, 1993; Son & Zhang, 1995; Son et al., 1996; Syberfeldt, 2007; Xiong et al., 2002) of RDRTDBS. System model proposed in (Pu & Leff, 1990; Son & Zhang, 1995; Gustavsson et al., 2004; Haj et al., 2008; Xiong et al., 2002) is a fully replicated type where data replica is replicated on all sites. This system model allows every admitted RTT to get processed locally. Although this system model is acceptable to satisfy the timeliness demand of DRTDBS, it suffers from the issue of unnecessary bandwidth utilization and unnecessary replica updation. Enforcing the cryptographic algorithm in such a system may ultimately increase the complexity of the system. Thus, we have explored another system model proposed in (Ulusoy, 1994; Kim, 1996; Peddi & DiPippo, 2002; Srivastava et al., 2012; Shrivastava & Shanker, 2018; Shrivastava & Shanker, 2019). In this system model, data replicas are partially replicated to some site. Although, this system overcomes the issue of full replication but in the majority of the time it causes short duration RTT to miss their deadline. The reason for missing it deadline is because of distributed processing of RTT and communication delay. Thus, we have not considered such a system model for enforcing a cryptographic algorithm. Apart from fully and partially replicated database, virtual full replication is proposed in (Mathiason et al., 2007). In this type of replicated system, data replicas are virtually fully replicated on all sites. Here, the updation of data replicas will depend on data requirement of RTT. Thus, based on demand requirement data replicas get updated. This type of data replication scheme solves the issues of fully and partially replicated system effectively and efficiently. Although, this replication type offers better service, but we have not considered such a system model for enforcing cryptographic algorithm because this replication type is an example of kernelized version and enforcing cryptographic algorithm may increase its complexity. This may ultimately cause RTT to either miss their deadline or to compromise in mutual consistency. A system model of (Salem et al., 2016) is also an example of virtual full replication. Recently, in the paper (Shrivastava & Shanker, 2018), we have found a system model where RPL operates from a middleware despite master sites or slave site. The master site and slave site only have to processes the update/write RTT and read RTT respectively. Due to the processing of RTT in their respected place location and shifting the complexity of maintaining mutual consistency in middleware, sites have limited request to process which ultimately increases the opportunity of admitted RTT to get completed within deadline. In this chapter, we have enforced the cryptographic algorithm in such a system model which secure the system from malicious attack. Specifically, we are securing the message which is propagating in the network in the plain text form. We have not considered the security within the system. The security within the system is provided by the antivirus.

Apart from enforcing the cryptographic algorithm, it is also important to decide which algorithm is to be used for encryption and decryption. There exists a different cryptographic algorithm (Padmavathi B. & Kumari, 2013) which encrypt/decrypt the message before transmitting in the network. These cryptographic algorithms can be categorized into symmetric key and asymmetric key algorithm. In these algorithms, key plays the main role to convert the plain text into cipher text and vice versa. Thus, in the cryptographic algorithm, it is important to secure the key from unauthorized user. The existing key which is used in the cryptographic is in the form of text which makes the task of attacker easy to crack the key. Thus, it is necessary to design the key in such a way that makes the task of an attacker very complex to crack the key. In a paper (Shrivastava et al., 2014), we have identified such a solution where the key is generated from the image and image is designed according to the user creativity. The key generation algorithm input the user designed image and extract the pixel value from such an image. This extracted pixel is XORed to generate the key. The generated key value is directly inputted to the DES algorithm to encrypt or decrypt the message. In this whole process of encryption and decryption, the user itself does not know the actual key which is passed in the DES. The user just knows about the image which is designed by him. Due to non-availability of key value, it is not possible to verbally transfer the value or crack the key value.

Overall, the issue of maintaining security in RDRTDBS is solved through enforcing cryptographic in the proposed system model (Shrivastava & Shanker, 2018) such that simultaneously mutual consistency, timeliness, and security can be provided.

## SOLUTIONS AND RECOMMENDATIONS

The normal processing of existing system model (Shrivastava & Shanker, 2018) with the cryptographic algorithm is illustrated in Figure 1.

In this system model, the main processing of maintaining mutual consistency is done by the middleware. Middleware consists of three main sub-layers (i.e. RTT data analyzer, conflict detection and resolution, and message propagation). This sub layers process different processing to maintain the mutual consistency between replicated data item. The algorithm code which works in the middleware is already proposed in (Shrivastava & Shanker, 2018) and interested authors are requested to read this paper. Apart from middleware, system model includes master sites, slave sites, and client. Master sites are a central place for the processing of update and write RTT

*Figure 1. Secure system mode of RDRTDBS*

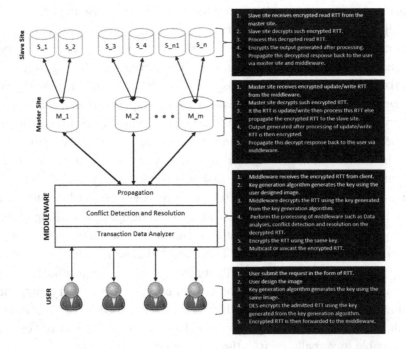

272

whereas slave sites process only read RTT. The client is the real service user who submits the request in the form of RTT to the middleware and middleware forward this admitted request to the master site or slave site. After processing of RTT in their respected site, the output is generated and is sent back to the client via middleware. All the messages that are exchanged in this system model for maintaining mutual consistency are propagated in the form of plain text. This plain text provides an easy opportunity to a hijacker to easily bypass the confidentiality and data integrity constraint of the database and cause malicious attack. Thus, to prevent the hijacker from accessing the plain text, we have enforced the cryptographic algorithm in such a system model. Cryptographic algorithm consists of two-part encryption/decryption algorithm and key. Algorithm is publicly available and is known to everyone. However, the key used in such an algorithm is made secure and is known to only intended sender or receiver. Thus, in order to make the cryptographic algorithm more secure, research must be conducted in the direction to make the key more secure. Recently, we have surveyed the paper (Shrivastava et al., 2014) and found a solution where research is conducted to make the key more secure. In the current book chapter, we have integrated such a solution in the system model proposed in (Shrivastava & Shanker, 2018).

To secure the admitted RTT and response from malicious attack in the system model (Shrivastava & Shanker, 2018), we have used the key generation algorithm proposed in (Shrivastava et al., 2014). In this algorithm, primarily user has to design the image according to its creativity. To design the image, user is provided with the user interface where user can design the image using in built controls. This in-built control includes 2D figures such as rectangle, square, circle, line and soon. Additionally, user interface consists of color chooser which allows the user to design the image using any color. This user interface is implemented in the java programming language and our extended system model is also implemented in the java language which makes the integration of both the mentioned approaches very easier. After designing the image, the output is saved in the format of jpeg. This output image is then forwarded as an input to the key generation algorithm proposed in (Shrivastava et al., 2014). Key generation algorithm extracts the pixel value from the image and generates the key which is again input in the DES algorithm. DES algorithm encrypts the admitted RTT using the same key. The algorithm code to encrypt the user message is also mentioned in (Shrivastava et al., 2014) and interested authors are requested to read the paper. After encryption of admitted RTT, encrypted RTT in the form of cipher text is then forwarded to the middleware. Additionally, image designed by the user for generating the key is also propagated to the middleware via asymmetric key cryptography. The flow diagram of such mentioned steps in shown in Figure 2.

*Figure 2. Encryption of admitted RTT*

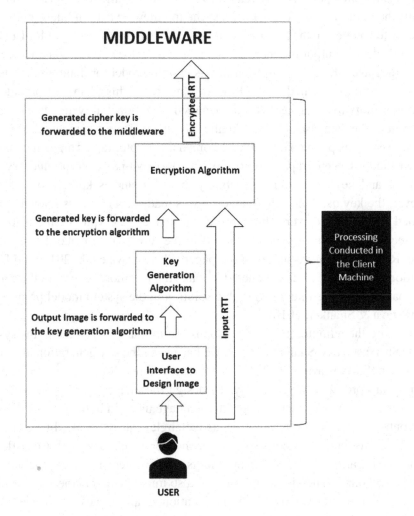

After reception of encrypted RTT and user designed image from the client machine, middleware uses the same image to generate the key from the key generation algorithm. After generation of the key, middleware decrypts the received encrypted RTT such that processing w.r.t to data analysis, conflict detection and resolution, and propagation get conducted. The flow diagram of such mentioned steps is shown in Figure 3. After processing in middleware, middleware encrypts the RTT and propagate to the master site to update or write the real time data item or non-real

*Figure 3. Decryption of admitted RTT*

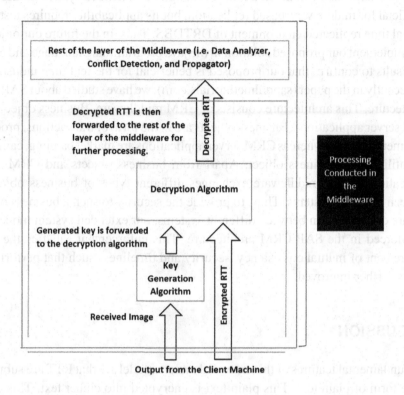

time data item respectively. master site or slave site performs the processing of admitted RTT and propagate back the response in an encrypted format to the user. Although enforcing cryptographic in a system model (Shrivastava & Shanker, 2018) increases the complexity, but it secures the system from the malicious attack which is important for future coming real time application.

The implementation of our extended system model with cryptographic algorithm demonstrate the importance of security in replicated environment of DRTDBS. Due to non-secured RDRTDBS, attackers can get easy access to any message transmitting in between client and master/slave sites. This gives an opportunity to the unintended user to read the confidential information and easily modifies the original content. This manipulated content is then forwarded by the attacker to the intended receiver to read this value. Receiver work on such modified information and generate inconsistent result that may cause heavy economical loss in the system. Our extended system avoids such occurrence and provides the security to the RTT from their admittance to the outcome. Although our extended system model looks beneficial for middleware-based replication, but its applicability requires testing in the real time replicated environment of DRTDBS. Thus, in the future our objective is to implement our proposed solution in the java programming language and collect the results to confirm that our proposed is beneficial for the real time industry.

Recently in the paper (saponlinetutorials.com), we have studied about SAP CRM architecture. This architecture consists of CRM middleware, Business objects, and CRM server applications. Business objects are in the form of transactions, products, customers and soon whereas CRM server applications are market panning, campaign that utilizes the business object. Apart from business objects and CRM server applications, CRM middleware exchanges different types of business objects in between different systems. Thus, to provide the security to such a business objects that are exchanging in between different systems, our extended system model may be enforced in the SAP CRM architecture. This enforcement satisfies the triple requirement of mutual consistency, security and timeliness such that performance can be further improved.

## DISCUSSION

The fundamental features of this extended system model are that RTT are submitted in the form of plain text. This plain text is encrypted into cipher text. This cipher text is then forwarded to the middleware. During the transmission of cipher in the network, attackers can possibly track the message. however, such encrypted message is no of use to the attacker because the key required to decrypt such message is generated from the user defined image and such image get exist only in the sender site and receiver site. Thus, cracking the key value is very complex for the attacker. Even though, if the attacker cracks the key, then such key value will be no use for the forth coming cipher text because user must design new image to generate new key for each newly admitted RTTs. Thus, cracking the key value for each coming cipher text is very complex for the attacker. More so, our extended system model (1) avoid the attacker from modifying the plain text; (2) avoids the attacker from accessing the actual image for generating the key value; (3) broadcasting the update/ write RTT in encrypted format to all the master sites at a common instance of time such that all master site get synchronously reach to the consistent state.

In addition to this, the performance of our proposed extended system model depends on the transaction miss ratio (TMR). TMR is the number of RTT that missed their deadline from the total number of RTTs. Thus, the primary performance metric of our extended system model is measured in terms of TMR. Additionally, the performance of secure RDRTDBS is measured in terms of RTTs that have successfully completed within their deadline following security constraint. Thus, another performance metric termed as transaction completed following security policy (i.e. TCSP) is used to measure the performance of secure RDRTDBS. Despite of TMR and TCSP, throughput is the third performance metric used to measure the performance of the system. Overall, the performance of our proposed extended system model is measured via TMR, TCSP, and Throughput.

## FUTURE RESEARCH DIRECTIONS

Apart from this cryptographic solution which we have discussed in this chapter, there is still more area that needs research.

1.  As already mentioned, that in existing cryptographic technique 2D image is used to generate the key. In the future, research can be conducted to use the 3D image to generate the key.
2.  Apart from the 3D graphics image, research can also be conducted to record the user's voice and generate the key from such digitized voice data.
3.  Research can be also conducted in the direction to record video of the user and generate the key from digitized video data.
4.  Research can be conducted to optimize the key generation algorithm such that performance of the secure RDRTDBS can be further improved.
5.  Research can be conducted to decrease the size of cipher text such that bandwidth utilization for transmitting the cipher can be overcome.
6.  Research can be conducted to decrease the size of image which is used to generate the key such that user designed image can be easily shared in between sender and receiver.
7.  Apart from using DES algorithm for encryption and decryption, research can be conducted to use more efficient algorithm such as AES, IDEAS and soon.

## CONCLUSION

The use of RT in DRTDBS is crucial to the many real time applications. However, when such replicated DRTDBS is utilized in securely demand real time application, a serious issue arises, i.e. how to maintain the confidentiality and preserve the data integrity of the replicated data item with maintaining mutual consistency and timeliness. RPL and system model play a vital role in maintaining the mutual consistency of replicated data item and timeliness of RTT. There exists different system model that ensures the mutual consistency of replicated data item, but such models have a large number of loopholes that provide an easy opportunity to a hijacker or unauthorized user to get bypass the security constraint and cause serious harm in the system. In this chapter, we have extended the system model proposed in (Shrivastava & Shanker, 2018) with the cryptographic algorithm proposed in (Shrivastava et al., 2014). The central idea behind this is to secure the admitted RTT from its entrance to the reception of output in the network. This extended system model with a cryptographic algorithm will perform better in terms of timeliness, mutual consistency, and security compared to other non-secure system models.

# REFERENCES

Gupta, A. K. & Swaroop, V. (2018). Overload handling in replicated real time distributed databases. *International Journal of Applied Engineering Research, 13*(18), 13969-13977.

Gustavsson, S. & Andler, S. F. (2004). Real-time conflict management in replicated databases. *Proceedings of the Fourth Conference for the Promotion of Research in IT at New Universities and University Colleges in Sweden (PROMOTE IT 2004)*, Karlstad, Sweden, 2. pp. 504-513. Academic Press.

Gustavsson, S., & Andler, S. R. (2005, April). Continuous consistency management in distributed real-time databases with multiple writers of replicated data. *Proceedings of the 19th IEEE International Parallel and Distributed Processing Symposium*. IEEE. 10.1109/IPDPS.2005.152

Haj Said, A., Sadeg, B., Amanton, L., & Ayeb, B. (2008). A protocol to control replication in distributed real-time database systems. *Proceedings of the Tenth International Conference on Enterprise Information Systems, 1*, pp. 501-504. Academic Press.

Kim, Y. K. (1996). *Towards real-time performance in a scalable, continuously available telecom. Redwood City, CA: DBMS.*

Lokhande, D. B., & Dhainje, P. B. (2019). A novel approach for transaction management in heterogeneous distributed real time replicated database systems. *International Journal for Scientific Research and Development, 7*(1), 840–844.

Mathiason, G., Andler, S. F., & Son, S. H. (2007, August). Virtual full replication by adaptive segmentation. *Proceedings of the 13th IEEE International Conference on Embedded and Real-Time Computing Systems and Applications (RTCSA 2007)* (pp. 327-336). IEEE.

Nie, T., & Zhang, T. (2009, January). A study of DES and Blowfish encryption algorithm. *Proceedings of the Tencon 2009-2009 IEEE Region 10 Conference* (pp. 1-4). IEEE.

Ouzzani, M., Medjahed, B., & Elmagarmid, A. K. (2009). Correctness criteria beyond serializability. In *Encyclopedia of database systems* (pp. 501–506). Academic Press.

Padmavathi, B., & Kumari, S. R. (2013). *A survey on performance analysis of DES, AES and RSA algorithm along with LSB substitution*. India: IJSR.

Peddi, P., & DiPippo, L. C. (2002). A replication strategy for distributed real-time object-oriented databases. *Proceedings Fifth IEEE International Symposium on Object-Oriented Real-Time Distributed Computing. ISIRC 2002* (pp. 129-136). IEEE. 10.1109/ISORC.2002.1003670

Pu, C. & Leff, A. (1990). Replica control in distributed systems: An asynchronous approach.

Salem, R., Saleh, S. A., & Abdul-Kader, H. (2016). Scalable data-oriented replication with flexible consistency in real-time data systems. *Data Science Journal*, 15.

saponlinetutorials.com. (n.d.). What is SAP CRM Middleware. Retrieved from https://www.saponlinetutorials.com/sap-crm-middleware/

Shanker, U., Misra, M., & Sarje, A. K. (2008). Distributed real time database systems: Background and literature review. *Distributed and Parallel Databases*, *23*(2), 127–149. doi:10.100710619-008-7024-5

Shrivastava, P., Jain, R., & Raghuwanshi, K. S. (2014, January). A modified approach of key manipulation in cryptography using 2d graphics image. *Proceedings of the 2014 International Conference on Electronic Systems, Signal Processing and Computing Technologies* (pp. 194-197). IEEE. 10.1109/ICESC.2014.40

Shrivastava, P. & Shanker, U. (2018, August). Replica update technique in RDRTDBS: Issues & challenges. *Proceedings of the 24th International Conference on Advanced Computing and Communications (ADCOM-2018)*, Bangalore, India (pp. 21-23). Academic Press.

Shrivastava, P., & Shanker, U. (2018). Replica control following 1SR in DRTDBS through best case of transaction execution. In *Advances in Data and Information Sciences* (pp. 139–150). Singapore: Springer. doi:10.1007/978-981-10-8360-0_13

Shrivastava, P., & Shanker, U. (2018). Replication protocol based on dynamic versioning of data object for replicated DRTDBS. *International Journal of Computational Intelligence & IoT*, *1*(2).

Shrivastava, P., & Shanker, U. (2019, January). Real time transaction management in replicated DRTDBS. *Proceedings of the Australasian Database Conference* (pp. 91-103). Springer. 10.1007/978-3-030-12079-5_7

Shrivastava, P., & Shanker, U. (2019, January). Supporting transaction predictability in replicated DRTDBS. *Proceedings of the International Conference on Distributed Computing and Internet Technology* (pp. 125-140). Springer. 10.1007/978-3-030-05366-6_10

Son, S. H., & Kouloumbis, S. (1993). A token-based synchronization scheme for distributed real-time databases. *Information Systems*, *18*(6), 375–389. doi:10.1016/0306-4379(93)90014-R

Son, S. H. & Zhang, F. (1995, April). Real-time replication control for distributed database systems: Algorithms and their performance. In DASFAA, 11, pp. 214-221.

Son, S. H., Zhang, F., & Hwang, B. (1996). Concurrency control for replicated data in distributed real-time systems. *Journal of Database Management*, *7*(2), 12–23. doi:10.4018/jdm.1996040102

Srivastava, A., Shankar, U., & Tiwari, S. K. (2012). A protocol for concurrency control in real-time replicated databases system. *International Journal of Computer Networks and Wireless Communications, 2*(3).

Syberfeldt, S. (2007). *Optimistic replication with forward conflict resolution in distributed real-time databases* [Doctoral dissertation]. Institutionen för datavetenskap.

Ulusoy, Ö. (1994). Processing real-time transactions in a replicated database system. *Distributed and Parallel Databases*, *2*(4), 405–436. doi:10.1007/BF01265321

Wang, F., Yao, L. W., & Yang, Y. L. (2011). Efficient verification of distributed real-time systems with broadcasting behaviors. *Real-Time Systems*, *47*(4), 285–318. doi:10.100711241-011-9122-0

Xiong, M., Ramamritham, K., Haritsa, J. R., & Stankovic, J. A. (2002). MIRROR: A state-conscious concurrency control protocol for replicated real-time databases. *Information Systems*, *27*(4), 277–297. doi:10.1016/S0306-4379(01)00053-9

Zhang, H., Zhao, W., Moser, L. E., & Melliar-Smith, P. M. (2011). Design and implementation of a byzantine fault tolerance framework for non-deterministic applications. *IET Software*, *5*(3), 342–356. doi:10.1049/iet-sen.2010.0013

Zhao, W. (2014). *Building dependable distributed systems*. John Wiley & Sons. doi:10.1002/9781118912744

# Compilation of References

Aberer, K., Hauswirth, M., & Salehi, A. (2006). *The Global Sensor Networks middleware for efficient and flexible deployment and interconnection of sensor networks* (No. REP_WORK).

Abubakar, A., & Pranggono, B. (2017, September). Machine learning based intrusion detection system for software defined networks. In *2017 Seventh International Conference on Emerging Security Technologies (EST)* (pp. 138-143). IEEE. 10.1109/EST.2017.8090413

Adeli, H., & Hung, S. L. (1994). *Machine learning: neural networks, genetic algorithms, and fuzzy systems*. Hoboken, NJ: John Wiley & Sons.

Adjih, C., Raffo, D., & Muhlethaler, P. (2005). Attacks against olsr: Distributed key management for security. *2nd OLSR Interop/Workshop,* Palaiseau, France, 14, pp. 1-5.

Ahmed, M. E., Kim, H., & Park, M. (2017, October). Mitigating DNS query-based DDoS attacks with machine learning on Software-defined networking. In 2017 IEEE Military Communications Conference (MILCOM) (pp. 11-16). IEEE. doi:10.1109/MILCOM.2017.8170802

Ahmed, E., Imran, M., Guizani, M., Rayes, A., Lloret, J., Han, G., & Guibene, W. (2017). Enabling mobile and wireless technologies for smart cities. *IEEE Communications Magazine, 55*(1), 74–75. doi:10.1109/MCOM.2017.7823341

Ahmed, E., Yaqoob, I., Hashem, I. A. T., Khan, I., Ahmed, A. I. A., Imran, M., & Vasilakos, A. V. (2017). The role of big data analytics in Internet of Things. *Computer Networks, 129*, 459–471. doi:10.1016/j.comnet.2017.06.013

Airehrour, D., Gutierrez, J., & Ray, S. K. (2016). Secure routing for internet of things: A survey. *Journal of Network and Computer Applications, 66*, 198–213. doi:10.1016/j.jnca.2016.03.006

Akyildiz, I. F., Wang, X., & Wang, W. (2005). Wireless Mesh Networks: A survey. *Computer Networks, 47*(4), 445–487. doi:10.1016/j.comnet.2004.12.001

Akyildiz, I., & Wang, X. (2009). *Wireless Mesh Networks (Advanced Texts in Communications and Networking)*. Chichester, UK: John Wiley & Sons.

Alajmi, N. M., & Elleithy, K. M. (2015). Comparative analysis of selective forwarding attacks over Wireless Sensor Networks. *International Journal of Computers and Applications, 111*(14).

Alam, T. (2019). Middleware Implementation in Cloud-MANET Mobility Model for Internet of Smart Devices. *arXiv preprint arXiv:1902.09744.*

Al-Janabi, S., Al-Shourbaji, I., Shojafar, M., & Shamshirband, S. (2017). Survey of main challenges (security and privacy) in wireless body area networks for healthcare applications. *Egyptian Informatics Journal, 18*(2), 113–122. doi:10.1016/j.eij.2016.11.001

Althumali, H., & Othman, M. (2018). A survey of random access control techniques for machine-to-machine communications in LTE/LTE-A networks. *IEEE Access: Practical Innovations, Open Solutions, 6*, 74961–74983. doi:10.1109/ACCESS.2018.2883440

Amer, W., Ansari, U., & Ghafoor, A. (2009). Industrial automation using embedded systems and machine-to-machine, man-to-machine (m2m) connectivity for improved overall equipment effectiveness (OEE). In *Proceedings of 2009 IEEE International Conference on Systems, Man and Cybernetics,* pp. 4450-4454.

Annunizata, M. & Bell, G. (2016). *Digital Future of the Electricity and Power Industry*, GE Power Digital Solutions, GE

Ansel, V. P., Aboothahir, M. A., Smritilakshmi, A. S., & Jose, B. R. (2016, December). Secure opportunistic large array for Internet of Things. In *Proceedings 2016 Sixth International Symposium on Embedded Computing and System Design (ISED)* (pp. 201-204). IEEE. 10.1109/ISED.2016.7977082

Asherson, S. & Hutchison, A. (n.d.). Secure routing in wireless mesh networks, *University of Cape Town*. Retrieved from http://pubs.cs.uct.ac.za/archive/00000318/01/SATNAC2006WIP.pdf

Atzori, L., Iera, A., & Morabito, G. (2010). The internet of things: A survey. *Computer Networks, 54*(15), 2787–2805. doi:10.1016/j.comnet.2010.05.010

Azmi, N. A., Samsul, S., Yamada, Y., Yakub, M. F. M., Ismail, M. I. M., & Dziyauddin, R. A. (2018, July). A survey of localization using RSSI and TDoA techniques in wireless sensor network: System architecture. In IEEE *2nd International Conference on Telematics and Future Generation Networks (TAFGEN)* (pp. 131-136).

Ballarini, P., Mokdad, L., & Monnet, Q. (2013). Modeling tools for detecting DoS attacks in WSNs. *Security and Communication Networks, 6*(4), 420–436. doi:10.1002ec.630

Bandiera, F., Coluccia, A., & Ricci, G. (2015). A cognitive algorithm for received signal strength-based localization. *IEEE Transactions on Signal Processing, 63*(7), 1726–1736. doi:10.1109/TSP.2015.2398839

Bankovic, Z., Fraga, D., Manuel, M. J., Carlos, V. J., Malagon, P., Araujo, A., & Nieto-Taladriz, O. (2010). Improving security in WMNs with reputation systems and self-organizing maps. *Journal of Network and Computer Applications*, *34*(2), 455–463. doi:10.1016/j.jnca.2010.03.023

Bankovic, Z., Vallejo, J. C., Fraga, D., & Moya, J. M. (2011). *Detecting bad-mouthing attacks on reputation systems using self-organizing maps. Computational Intelligence in Security for Information Systems* (pp. 9–16). Springer.

Bansal, D., Sofat, S., Pathak, P., & Bhoot, S. (2011). Detecting MAC misbehaviour switching attacks in Wireless Mesh Networks. *International Journal of Computers and Applications*, *26*(5), 55–62. doi:10.5120/3102-4261

Barki, L., Shidling, A., Meti, N., Narayan, D. G., & Mulla, M. M. (2016, September). Detection of distributed denial of service attacks in software defined networks. In *2016 International Conference on Advances in Computing, Communications and Informatics (ICACCI)* (pp. 2576-2581). IEEE. 10.1109/ICACCI.2016.7732445

Barnaghi, P. (2010). Sensei: An architecture for the real world internet. In *First International Workshop on Semantic Interoperablility for Smart Spaces.*

Behera, G., Panigrahy, S. K., & Turuk, A. K. (2018, February). A biometric based anonymous user authentication technique in wireless body area networks. In *2018 International Conference on Communication, Computing and Internet of Things (IC3IoT)* (pp. 308-312). IEEE. 10.1109/IC3IoT.2018.8668193

Ben-Othman, J., & Benitez, Y. I. S. (2011). IBC-HWMP: A novel secure identity-based cryptography-based scheme for Hybrid Wireless Mesh Protocol for IEEE 802.11s. *Concurrency and Computation*, *25*(5), 686–700. doi:10.1002/cpe.1813

Ben-Othman, J., & Benitez, Y. I. S. On securing HWMP using IBC, *IEEE International Conference on Communications*, *Kyoto, Japan*, 1–5. doi:10.1109/icc.2011.5962921

Bentov, I., Lee, C., Mizrahi, A., & Rosenfeld, M. (2014, June). Proof of activity: Extending bitcoin's proof of work via proof of stake. In *Proceedings of the 9th Workshop on the Economics of Networks, Systems and Computation*, Austin, TX.

Biswas, D., Ramamurthy, R., Edward, S. P., & Dixit, A. (2015). The Internet of Things: Impact and applications in the high-tech industry, Cognizant 20-20 Insights, Cognizant Technology Services, New Jersey. Available at https://www. cognizant. com/whitepapers/the-internet-of-things-impact-and-applicationsin-the-high-tech-industry-codex1223. pdf.

Bletsas, A., Khisti, A., Reed, D. P., & Lippman, A. (2005). A simple cooperative diversity method based on network path selection. *arXiv preprint cs/0510071*.

Blockchain of things. (2017). Available online: https://www.blockchainofthings.com (*Accessed 1 February 2018*)

Boldrini, C., Conti, M., Delmastro, F., & Passarella, A. (2010). Context-and social-aware middleware for opportunistic networks. *Journal of Network and Computer Applications, 33*(5), 525–541. doi:10.1016/j.jnca.2010.03.017

Borge, M., Kokoris-Kogias, E., Jovanovic, P., Gasser, L., Gailly, N., & Ford, B. (2017, April). Proof-of-personhood: Redemocratizing permissionless cryptocurrencies. In *Proceedings of the IEEE European Symposium on Security and Privacy Workshop*, Paris, France.

Boudguiga, A., & Laurent, M. (2012). An authentications scheme for IEEE802.11s mesh networks relying on Sakai-Kasahara ID-Based Cryptographic Algorithms, *International Conference on Communications and Networking, Niagara Falls*, 256–263. doi:10.1109/comnet.2012.6217728

Boukerche, A., Oliveira, H. A., Nakamura, E. F., & Loureiro, A. A. (2007). Localization systems for wireless sensor networks. *IEEE Wireless Communications, 14*(6), 6–12. doi:10.1109/MWC.2007.4407221

Boyes, H., Hallaq, B., Cunningham, J., & Watson, T. (2018). The industrial internet of things (IIoT): An analysis framework. *Computers in Industry, 101*, 1–12. doi:10.1016/j.compind.2018.04.015

Brachmann, M., Garcia-Morchon, O., Keoh, S.-L., & Kumar, S. S. (2012). Security considerations around end-to-end security in the IP-based internet of things. In *Proceedings 2012 Workshop on Smart Object Security, in Conjunction with IETF83*, pp. 1–3.

Cachin, C., Schubert, S., & Vukolic, M. (2016, December). Non-determinism in Byzantine´ fault-tolerant replication. In *Proceedings of the International Conference on Principles of Distributed Systems (OPODIS 2016)*, Madrid, Spain.

Cadger, F., Curran, K., Santos, J., & Moffett, S. (2013). A survey of geographical routing in wireless ad-hoc networks. *IEEE Communications Surveys and Tutorials, 15*(2), 621–653. doi:10.1109/SURV.2012.062612.00109

Calyptix. (2015). DDoS Attacks 101: Types, targets, and motivations. Retrieved from https://www.calyptix.com/top-threats/ddos-attacks-101-types-targets-motivations/

Caporuscio, M., Raverdy, P. G., & Issarny, V. (2012). ubiSOAP: A service-oriented middleware for ubiquitous networking. *IEEE Transactions on Services Computing, 5*(1), 86–98. doi:10.1109/TSC.2010.60

Castelluccia, C., Chan, A. C., Mykletun, E., & Tsudik, G. (2009). Efficient and provably secure aggregation of encrypted data in wireless sensor networks. [TOSN]. *ACM Transactions on Sensor Networks, 5*(3), 20. doi:10.1145/1525856.1525858

Castro, M. & Liskov, B. (1999, February). Practical Byzantine fault tolerance. In *Proceedings of the Third Symposium on Operating Systems Design and Implementation,* New Orleans, LA.

Chan, H., Perrig, A., & Song, D. (2003, May). Random key predistribution schemes for sensor networks. In IEEE Symposium on Security and Privacy, Oakland, CA, (pp. 197-213).

Chelouah, L., Semchedine, F., & Bouallouche-Medjkoune, L. (2018). Localization protocols for mobile wireless sensor networks: A survey. *Computers & Electrical Engineering, 71,* 733–751. doi:10.1016/j.compeleceng.2017.03.024

Chen, T., Kuo, G.-S., Li, Z.-P., & Zhu, G. M. (2009). Intrusion detection in Wireless Mesh Networks Security, Wireless Mesh Networks. Boca Raton, FL: Auerbach Publications.

Chen, Z., Chen, S., & Feng, X. (2016, October). A design of distributed storage and processing system for internet of vehicles. In *Proceedings 2016 8th International Conference on Wireless Communications & Signal Processing (WCSP)* (pp. 1-5). IEEE. 10.1109/WCSP.2016.7752671

Chen, C. M., Lin, Y. H., Lin, Y. C., & Sun, H. M. (2012). RCDA: Recoverable concealed data aggregation for data integrity in wireless sensor networks. *IEEE Transactions on Parallel and Distributed Systems, 23*(4), 727–734. doi:10.1109/TPDS.2011.219

Chen, K., & Shen, H. (2017). FaceChange: Attaining neighbor node anonymity in mobile opportunistic social networks with fine-grained control. [TON]. *IEEE/ACM Transactions on Networking, 25*(2), 1176–1189. doi:10.1109/TNET.2016.2623521

Chen, L., Thombre, S., Järvinen, K., Lohan, E. S., Alén-Savikko, A., Leppäkoski, H., ... Lindqvist, J. (2017). Robustness, security and privacy in location-based services for future IoT: A survey. *IEEE Access: Practical Innovations, Open Solutions, 5,* 8956–8977. doi:10.1109/ACCESS.2017.2695525

Chen, R., Guo, J., & Bao, F. (2016). Trust management for SOA-based IoT and its application to service composition. *IEEE Transactions on Services Computing, 9*(3), 482–495. doi:10.1109/TSC.2014.2365797

Chen, Z., Jiang, F., Cheng, Y., Gu, X., Liu, W., & Peng, J. (2018, January). Xgboost Classifier for DDoS Attack Detection and Analysis in SDN-Based Cloud. In *2018 IEEE International Conference on Big Data and Smart Computing (BigComp)* (pp. 251-256). IEEE. 10.1109/BigComp.2018.00044

Chhabra, A., Vashishth, V., & Sharma, D. K. (2017, March). A game theory based secure model against Black hole attacks in Opportunistic Networks. In *Proceedings 2017 51st Annual Conference on Information Sciences and Systems (CISS)* (pp. 1-6). IEEE. 10.1109/CISS.2017.7926114

Chitkara, R., & Mesirow, R. (2017). *The industrial internet of things.* London, UK: PricewaterhouseCoopers.

Choudhary, G., Sharma, V., You, I., Yim, K., Chen, I. R., & Cho, J. H. (2018). Intrusion detection systems for networked unmanned aerial vehicles: A survey. *Proceedings of the 14th International Wireless Communications & Mobile Computing Conference, Limassol, Cyprus*, 560-565.

Chowdhury, T. J., Elkin, C., Devabhaktuni, V., Rawat, D. B., & Oluoch, J. (2016). Advances on localization techniques for wireless sensor networks: A survey. *Computer Networks*, *110*, 284–305. doi:10.1016/j.comnet.2016.10.006

Colla, M., Leidi, T., & Semo, M. (2009), Design and implementation of industrial automation control systems: A survey. In *Proceedings of 7th IEEE International Conference on Industrial Informatics* 10.1109/INDIN.2009.5195866

Consel, C. & Kabac, M. (2017). Internet of things: From small- to large-scale orchestration. In *IEEE 37th International Conference on Distributed Computing Systems*.

Conway, J. (2015). The industrial internet of things: An evolution to a smart manufacturing enterprise. *Schneider Electric*.

Conzon, D., Bolognesi, T., Brizzi, P., Lotito, A., Tomasi, R., & Spirito, M. A. (2012). The VIRTUS middleware: An XMPP based architecture for secure IoT communications. In *Proceedings 2012 21st International Conference on Computer Communications and Networks, ICCCN*. doi:.6289309. pp 1-6.10.1109/ICCCN.2012

Copeland, C. & Zhong, H. (2018). *Tangaroa: A Byzantine fault tolerant raft*. Available at http://www.scs.stanford.edu/14au- cs244b/labs/projects/copeland_zhong.pdf

Cotton, S. L. & Scanlon, W. G. (2006, September). A statistical analysis of indoor multipath fading for a narrowband wireless body area network. In *2006 IEEE 17th International Symposium on Personal, Indoor, and Mobile Radio Communications.* (pp. 1-5).

da Silva, A. S., Wickboldt, J. A., Granville, L. Z., & Schaeffer-Filho, A. (2016, April). ATLANTIC: A framework for anomaly traffic detection, classification, and mitigation in SDN. In *Proceedings 2016 IEEE/IFIP Network Operations and Management Symposium (NOMS)* (pp. 27-35). IEEE. 10.1109/NOMS.2016.7502793

Deb, N., Chakraborty, M., & Chaki, N. (2011). A state-of-the-art survey on Ids for mobile ad-hoc networks and wireless mesh networks. *Communications in Computer and Information Science*, *203*, 169–179. doi:10.1007/978-3-642-24037-9_17

Demirbas, M. & Song, Y. (2006). An RSSI-based scheme for sybil attack detection in wireless sensor networks. In *Proceedings of the 2006 International Symposium on World of Wireless, Mobile and Multimedia Networks, WOWMOM '06*, IEEE Computer Society, Washington, DC, pp. 564–570. 10.1109/WOWMOM.2006.27

Dhanjani, N. (2015). Abusing the Internet of Things blackouts. *Freakouts, and Stakeouts*. O'Relly Media.

Dhurandher, S. K., Kumar, A., Woungang, I., & Obaidat, M. S. (2017). Supernova and hypernova misbehavior detection scheme for opportunistic networks. IEEE 31st International Conference on Advanced Information Networking and Applications (AINA). IEEE, pp. 387–391. 10.1109/AINA.2017.17

Dhurandher, S. K., Kumar, A., & Obaidat, M. S. (2017). Cryptography-based misbehavior detection and trust control mechanism for opportunistic network systems. *IEEE Systems Journal*, (99), 1–12.

Dhurandher, S. K., Woungang, I., Arora, J., & Gupta, H. (2016). History-based secure routing protocol to detect blackhole and greyhole attacks in opportunistic networks. [Formerly Recent Patents on Telecommunication]. *Recent Advances in Communications and Networking Technology*, *5*(2), 73–89.

Distl, B. & Neuhaus, S. (2015, January). Social power for privacy protected opportunistic networks. In *2015 7th International Conference on Communication Systems and Networks (COMSNETS)* (pp. 1-8). IEEE. 10.1109/COMSNETS.2015.7098697

Duan, J., Gao, D., Yang, D., Foh, C. H., & Chen, H. H. (2014). An energy-aware trust derivation scheme with game theoretic approach in wireless sensor networks for IoT applications. *IEEE Internet of Things Journal*, *1*(1), 58–69. doi:10.1109/JIOT.2014.2314132

Durahim, A. O., & Savaş, E. (2010). A-MAKE: An efficient, anonymous and accountable authentication framework for WMNs. In *Proceedings of 5th International Conference on Internet Monitoring and Protection, Barcelona, Spain*, 54-59. 10.1109/ICIMP.2010.16

Dvir, A., Holczer, T., & Buttyan, L. (2011). VeRA - version number and rank authentication in RPL. In *2011 IEEE Eighth International Conference on Mobile Ad-Hoc and Sensor Systems*. pp. 709-714.

Egners, A., & Meyer, U. (2010). Wireless Mesh Network Security: State of Affairs. In *Proceedings of 6th IEEE Workshop on Security in Communication Networks, Denver, Colorado*, 997-1004, 10.1109/LCN.2010.5735848

Eisenhauer, M., Rosengren, P., & Antolin, P. (2010). Hydra: A development platform for integrating wireless devices and sensors into ambient intelligence systems. In *The Internet of Things* (pp. 367–373). New York, NY: Springer. doi:10.1007/978-1-4419-1674-7_36

Eschenauer, L., & Gligor, V. D. (2002, November). A key-management scheme for distributed sensor networks. In *Proceedings of the 9th ACM conference on Computer and communications security* (pp. 41-47). ACM. 10.1145/586110.586117

Ester, M., Kriegel, H. P., Sander, J., & Xu, X. (1996, August). A density-based algorithm for discovering clusters in large spatial databases with noise. *Proceedings of International Conference on Knowledge Discovery & Data Mining (KDD)*, *96*(34), 226–231.

Evans, P. C. & Annunziata, M. (2012, November). *Industrial internet: Pushing the boundaries of minds and machines*, GE Digital, GE.

Evgeniou, T. & Pontil, M. (2001). Workshop on support vector machines: Theory and applications.

Eyal, I., Gencer, A. E., Sirer, E. G., & Van Renesse, R. (2016, March). Bitcoin-NG: A scalable blockchain protocol. In *Proceedings of the 13th USENIX Symposium on Networked Systems Design and Implementation*, Santa Clara, CA.

Eyal, I., Gencer, A. E., Sirer, E. G., & Van Renesse, R. (2016). Bitcoin-NG: a scalable blockchain protocol. In *Proceedings 13th USENIX Symposium on Networked Systems Design and Implementation (NSDI 16)*, Santa Clara, CA, pp. 45–59.

Fadel, E., Gungor, V. C., Nassef, L., Akkari, N., Malik, M. A., Almasri, S., & Akyildiz, I. F. (2015). A survey on wireless sensor networks for smart grid. *Computer Communications*, *71*, 22–33. doi:10.1016/j.comcom.2015.09.006

Fall, K. (2003). A delay-tolerant network architecture for challenged internets. *Proceedings of the 2003 Conference on Applications, Technologies, Architectures, and Protocols for Computer Communications*. ACM, pp. 27–34. 10.1145/863955.863960

Feng, Y., Fan, M. Y., & Liu, C. P. (2008). A new privacy-enhanced authentication scheme for Wireless Mesh Networks. *School of Computer Science and Technology. University of Electronic Science and Technology of China*, *265-269*. doi:10.1109/ICACIA.2008.4770020

Ferdowsi, A. & Saad, W. (2017). Deep learning-based dynamic watermarking for secure signal authentication in the internet of things.

Fernandez-Gago, C., Moyano, F., & Lopez, J. (2017). Modelling trust dynamics in the internet of things. *Information Science*, *396*, 72–82. doi:10.1016/j.ins.2017.02.039

Fiege, E., Hammer, M., Ulrich, R., Iacovelli, D., & Bromberger, J. (2016). *Industry 4.0 at McKinsey's model factories*. McKinsey Corporation GE, Digital Rig, Retrieved from https://www.ge.com/digital/blog/worlds-first-digital-rig-digitizing-operational-excellence

Fields, R. E. (1997). Evaluating compliance with FCC guidelines for human exposure to radiofrequency electromagnetic fields. *OET bulletin, 65*(10).

Filament. (2017). Available at https://filament.com/

Fogal, D., Rauscheker, U., Lanctot, P., Bildstein, A., Burhop, M., Caneodo, A., ... Xiaonan, S. (2015). Factory of the future. *IEC*.

Fort, A., Desset, C., Wambacq, P., & Biesen, L. V. (2007). Indoor body-area channel model for narrowband communications. *IET Microwaves, Antennas & Propagation, 1*(6), 1197–1203. doi:10.1049/iet-map:20060215

Fujisaki, K. (2019). Evaluation of 13.56 MHz RFID system performance considering communication distance between reader and tag. *Journal of High Speed Networks, 25*(1), 61–71. doi:10.3233/JHS-190603

Fu, Y., He, J., Luan, L., Wang, R., & Li, G. (2008). A zone-based distributed key management scheme for Wireless Mesh Networks. In *Proceedings of 32nd Annual IEEE International Computer Software and Applications Conference, Turku, Finland*, 68–71. 10.1109/COMPSAC.2008.131

Fu, Y., He, J., Wang, R., & Li, G. (2008). Mutual authentication in Wireless Mesh Networks. In *Proceedings of IEEE International Conference on Communications*, Beijing, China, 1690–1694. doi:10.1109/icc.2008.326

Gandino, F., Montrucchio, B., & Rebaudengo, M. (2009, December). Random key pre-distribution with transitory master key for wireless sensor networks. In *Proceedings of the 5th International Student Workshop on Emerging Networking Experiments and Technologies* (pp. 27-28). ACM. 10.1145/1658997.1659012

Gandino, F., Montrucchio, B., & Rebaudengo, M. (2014). Key management for static wireless sensor networks with node adding. *IEEE Transactions on Industrial Informatics, 10*(2), 1133–1143. doi:10.1109/TII.2013.2288063

Gan, S. (2017). *An IoT Simulator in NS3 and a Key-Based Authentication Architecture for IoT Devices using Blockchain*. Indian Institute of Technology Kanpur.

Gao, L., Chang, E., Parvin, S., Han, S., & Dillon, T. (2010). A secure key management model for Wireless Mesh Networks. In *Proceedings of 24th IEEE International Conference on Advanced Information Networking and Applications, Perth, WA*, 655–660. 10.1109/AINA.2010.110

Gharghan, S. K., Nordin, R., Ismail, M., & Ali, J. A. (2016). Accurate wireless sensor localization technique based on hybrid PSO-ANN algorithm for indoor and outdoor track cycling. *IEEE Sensors Journal, 16*(2), 529–541. doi:10.1109/JSEN.2015.2483745

Glass, S. M., Muthukkumurasamy, V., & Portmann, M. (2009). Detecting man-in-the-middle and wormhole attacks in Wireless Mesh Networks. In *Proceedings of International Conference on Advanced Information Networking and Applications, Bradford, UK*, 530–538. 10.1109/AINA.2009.131

Glass, S., Portmann, M., & Muthukkumarasamy, V. (2008). Securing Wireless Mesh Networking. *IEEE Internet Computing, 12*(4), 30–36. doi:10.1109/MIC.2008.85

Gmez-Goiri, A., Ordua, P., Diego, J., & de Ipiña, D. L. (2014). *Otsopack: Lightweight semantic framework for interoperable ambient intelligence applications* (pp. 460–467). Computer Human Behaviour. doi:10.1016/j.chb

Goeckel, D., Vasudevan, S., Towsley, D., Adams, S., Ding, Z., & Leung, K. (2011). Artificial noise generation from cooperative relays for everlasting secrecy in two-hop wireless networks. *IEEE Journal on Selected Areas in Communications, 29*(10), 2067–2076. doi:10.1109/JSAC.2011.111216

Goel, S., & Negi, R. (2008). Guaranteeing secrecy using artificial noise. *IEEE Transactions on Wireless Communications, 7*(6), 2180–2189. doi:10.1109/TWC.2008.060848

Goodin, D. (2018). *US service provider survives the biggest recorded DDoS in history*. Retrieved from https://arstechnica.com/information-technology/2018/03/us-service-provider-survives-the-biggest-recorded-ddos-in-history/

Goransson, P., Black, C., & Culver, T. (2016). *Software defined networks: A comprehensive approach*. Burlington, MA: Morgan Kaufmann.

Granjal, J., Monteiro, E., & Silva, J. S. (2013). Application-layer security for the WoT: Extending CoAP to support end-to-end message security for internet-integrated sensing applications. In *Proceedings International Conference on Wired/Wireless Internet Communication, Springer Berlin Heidelberg*, pp. 140–153. 10.1007/978-3-642-38401-1_11

Guinard, D., Trifa, V., Karnouskos, S., Spiess, P., & Savio, D. (2010). Interacting with the soa-based internet of things: Discovery, query, selection, and on-demand provisioning of web services. *IEEE Transactions on Services Computing, 3*(3), 223–235. doi:10.1109/TSC.2010.3

Guo, B., Zhang, D., Wang, Z., Yu, Z., & Zhou, X. (2013). Opportunistic IoT: Exploring the harmonious interaction between human and the internet of things. *Journal of Network and Computer Applications, 36*(6), 1531–1539. doi:10.1016/j.jnca.2012.12.028

Guo, H., Wang, X., Cheng, H., & Huang, M. (2016). A routing defense mechanism using evolutionary game theory for delay tolerant networks. *Applied Soft Computing, 38*, 469–476. doi:10.1016/j.asoc.2015.10.019

Gupta, A. K. & Swaroop, V. (2018). Overload handling in replicated real time distributed databases. *International Journal of Applied Engineering Research, 13*(18), 13969-13977.

Gupta, B. B., Joshi, R. C., & Misra, M. (2009). Defending against distributed denial of service attacks: Issues and challenges. *Information Security Journal: A Global Perspective, 18*(5), 224-247.

Gupta, N. & Dayal, N. (2018, August). Optimal cache placement by identifying possible congestion points in wireless sensor networks. In *International Conference on Wireless Intelligent and Distributed Environment for Communication* (pp. 161-170). Cham, Switzerland: Springer. 10.1007/978-3-319-75626-4_12

Gupta, S., Dhurandher, S. K., Woungang, I., Kumar, A., & Obaidat, M. S. (2013, October). Trust-based security protocol against blackhole attacks in opportunistic networks. In *2013 IEEE 9th International Conference on Wireless and Mobile Computing, Networking and Communications (WiMob)* (pp. 724-729). IEEE. 10.1109/WiMOB.2013.6673436

Gustavsson, S. & Andler, S. F. (2004). Real-time conflict management in replicated databases. *Proceedings of the Fourth Conference for the Promotion of Research in IT at New Universities and University Colleges in Sweden (PROMOTE IT 2004)*, Karlstad, Sweden, 2, pp. 504-513. Academic Press.

Gustavsson, S., & Andler, S. R. (2005, April). Continuous consistency management in distributed real-time databases with multiple writers of replicated data. *Proceedings of the 19th IEEE International Parallel and Distributed Processing Symposium*. IEEE. 10.1109/IPDPS.2005.152

Haj Said, A., Sadeg, B., Amanton, L., & Ayeb, B. (2008). A protocol to control replication in distributed real-time database systems. *Proceedings of the Tenth International Conference on Enterprise Information Systems, 1*, pp. 501-504. Academic Press.

Hakiri, A., Gokhale, A., Berthou, P., Schmidt, D. C., & Gayraud, T. (2014). Software-defined networking: Challenges and research opportunities for future internet. *Computer Networks, 75*, 453–471. doi:10.1016/j.comnet.2014.10.015

Halder, S., & Ghosal, A. (2016). A survey on mobility-assisted localization techniques in wireless sensor networks. *Journal of Network and Computer Applications, 60*, 82–94. doi:10.1016/j.jnca.2015.11.019

Hamid, Z., & Khan, S. A. (2006). An Augmented Security Protocol for Wireless MAN Mesh Networks. In *Proceedings of International Symposium on Communications and Information Technologies*, 861-865. doi:10.1109/iscit.2006.339859

Han, B., Yang, X., Sun, Z., Huang, J., & Su, J. (2018). OverWatch: A cross-plane DDoS attack defense framework with collaborative intelligence in SDN. *Security and Communication Networks, 2018*, 1–15. doi:10.1155/2018/9649643

Han, G., Jiang, J., Zhang, C., Duong, T. Q., Guizani, M., & Karagiannidis, G. K. (2016). A survey on mobile anchor node assisted localization in wireless sensor networks. *IEEE Communications Surveys and Tutorials, 18*(3), 2220–2243. doi:10.1109/COMST.2016.2544751

Harris, K. *(2008, September). An application of IEEE 1588 to industrial automation. In 2008 IEEE International Symposium on Precision Clock Synchronization for Measurement, Control and Communication (pp. 71-76). IEEE.*

Hector, M., Guha, L. C., & Lu, R. K. (2011). A secure service-oriented routing algorithm for heterogeneous wireless mesh networks. In *Proceedings of IEEE Global Telecommunications Conference*, 1-5. doi: 10.1109/GLOCOM.2011.6134439

Hoffman, S. (2015). Apache flume: Distributed log collection for hadoop. Birmingham, UK: Packt Publishing.

Hsu, C. W., Chang, C. C., & Lin, C. J. (2003). A practical guide to support vector classification.

Hugelshofer, F., Smith, P., Hutchison, D., & Race, N. J. P. (2009). OpenLIDS: A lightweight intrusion detection system for wireless mesh networks. In *Proceedings of 15th Annual International Conference on Mobile Computing and Networking, China*, 309-320. 10.1145/1614320.1614355

Hu, Y.-C., Perrig, A., & Johnson, D. B. (2005). Ariadne: A secure on-demand routing protocol for ad hoc networks. *Wireless Networks, 11*(1), 21–38. doi:10.100711276-004-4744-y

Iqbal, M. H., & Soomro, T. R. (2015). Big data analysis: Apache storm perspective. *International Journal of Computer Trends and Technology, 19*(1), 9–14. doi:10.14445/22312803/IJCTT-V19P103

Islam, S., Hamid, A., & Hong, C. S. (2009). *SHWMP: a secure hybrid wireless mesh protocol for IEEE 802.11s wireless mesh networks. Transactions on Computational Science, 5730* (pp. 95–114). Berlin, Germany: Springer-Verlag.

Jiacheng, H., Ning, L., Ping, Y., Futai, Z., & Qiang, Z. (2010). Securing Wireless Mesh Network with mobile firewall. In *Proceedings of IEEE International Conference on Wireless Communications and Signal Processing*, Suzhou, China, 1–6. 10.1109/WCSP.2010.5633566

Jin, J., Gubbi, J., Luo, T., & Palaniswami, M. (2012, October). Network architecture and QoS issues in the internet of things for a smart city. In *2012 International Symposium on Communications and Information Technologies (ISCIT)*, (pp. 956-961). IEEE. 10.1109/ISCIT.2012.6381043

Kaisler, S., Armour, F., Espinosa, J. A., & Money, W. (2013, January). Big data: Issues and challenges moving forward. In *Proceedings 2013 46th Hawaii International Conference on System Sciences* (pp. 995-1004). IEEE.

Kaiza, J. (2016). A new age of industrial production The Internet of things, services and people. Zurich, Switzerland: ABB Group.

Kandhoul, N. & Dhurandher, S. K. (2019). An asymmetric RSA based security approach for opportunistic IoT. *Proceedings 2nd International Conference on Wireless, Intelligent and Distributed Environment for Communication*. Springer, WIDECOM, Milan, Italy, pp. 47–60.

Kapetanovic, D., Zheng, G., & Rusek, F. (2015). Physical layer security for massive mimo: An overview on passive eavesdropping and active attacks. *arXiv preprint arXiv:1504.07154*.

Kapil, G., Agrawal, A., & Khan, R. A. (2016, October). A study of big data characteristics. In *Proceedings 2016 International Conference on Communication and Electronics Systems (ICCES)* (pp. 1-4). IEEE.

Kaplantzis, S., Shilton, A., Mani, N., & Sekercioglu, Y. A. (2007). Detecting selective forwarding attacks in wireless sensor networks using support vector machines. *Proceedings of International Conference on Intelligent Sensors, Sensor Networking and Information*. 10.1109/ISSNIP.2007.4496866

Karlof, C., & Wagner, D. 2003). Secure routing in wireless sensor networks: Attacks and countermeasures. *Proceedings of the First IEEE International Workshop on Sensor Network Protocols and Applications*. IEEE, pp. 113–127. 10.1109/SNPA.2003.1203362

Karthik, N. & Dhulipala, V. S. (2011, April). Trust calculation in wireless sensor networks. In *2011 3rd International Conference on Electronics Computer Technology*, 4, pp. 376-380. IEEE. 10.1109/ICECTECH.2011.5941924

Katyara, S., Shah, M. A., Zardari, S., Chowdhry, B. S., & Kumar, W. (2017). WSN based smart control and remote field monitoring of Pakistan's irrigation system using SCADA applications. *Wireless Personal Communications*, 95(2), 491–504. doi:10.100711277-016-3905-5

Keramatpour, A., Nikanjam, A., & Ghaffarian, H. (2017). Deployment of wireless intrusion detection systems to provide the most possible coverage in wireless sensor networks without infrastructures. *Wireless Personal Communications*, 96(3), 3965–3978. doi:10.100711277-017-4363-4

Keränen, A., Ott, J., & Kärkkäinen, T. (2009, March). The ONE simulator for DTN protocol evaluation. In *Proceedings of the 2nd International Conference on Simulation Tools and Techniques* (p. 55). ICST (Institute for Computer Sciences, Social-Informatics and Telecommunications Engineering). 10.4108/ICST.SIMUTOOLS2009.5674

Khan, M. A. & Salah, K. (2018). IoT security: Review, blockchain solutions, and open challenges. *Future Generation Computer Systems*. pp. 395–411.

Khan, M., Wu, X., & Dou, W. (2017), Big Data Challenges and Opportunities in the Hype of Industry 4.0, Big Data Networking Track, *IEEE ICC 2017 SAC Symposium*

Khan, M. A., & Salah, K. (2017). *IoT security: Review, blockchain solutions, and open challenges. Future Generation Computer Systems*.

Khan, S., Leo, K. K., & Din, Z. U. (2010). Framework for intrusion detection in IEEE 802.11 Wireless Mesh Networks. *The International Arab Journal of Information Technology, 7*(4), 435–440.

Khan, S., Loo, K. K., Mast, N., & Naeem, T. (2010). SRPM: Secure routing protocol for IEEE802.11 infrastructure based wireless mesh networks. *Journal of Network and Systems Management, 18*(2), 190–209. doi:10.100710922-009-9143-3

Khan, S., Nabil, A. A., & Loo, K. K. (2012). Secure route selection in Wireless Mesh Networks. *Computer Networks, 56*(2), 491–503. doi:10.1016/j.comnet.2011.07.005

Khattak, Z. K., Awais, M., & Iqbal, A. (2014, December). Performance evaluation of OpenDaylight SDN controller. In *Proceedings 2014 20th IEEE International Conference on Parallel and Distributed Systems (ICPADS)* (pp. 671-676). IEEE. 10.1109/PADSW.2014.7097868

Khurshid, A., Khan, A. N., Khan, F. G., Ali, M., Shuja, J., & Khan, A. U. R. (2018). Secure-CamFlow: A device-oriented security model to assist information flow control systems in cloud environments for IoTs. *Concurrency and Computation*, e4729.

Kim, Y. K. (1996). *Towards real-time performance in a scalable, continuously available telecom.* Redwood City, CA: DBMS.

Kim, H. Protection against packet fragmentation attacks at 6LoWPAN adaptation layer. In *Proceedings 2008 International Conference on Convergence and Hybrid Information Technology.* . pp. 790-801. 10.1109/ICHIT.2008.261

Kokila, R. T., Selvi, S. T., & Govindarajan, K. (2014, December). DDoS detection and analysis in SDN-based environment using support vector machine classifier. In *Proceedings 2014 Sixth International Conference on Advanced Computing (ICoAC)* (pp. 205-210). IEEE. 10.1109/ICoAC.2014.7229711

Kothmayr, T., Schmitt, C., Hu, W., Brnig, M., & Carle, G. (2012). *37th annual IEEE Conference on Local Computer Networks - Workshops.* doi:. pp. 964–972. 10.1109/LCNW.2012.6424088

Krazit, T. (2018). *What are memcached servers, and why are they being used to launch record-setting DDoS attacks?* Retrieved from https://www.geekwire.com/2018/memcached-servers-used-launch-record-setting-ddos-attacks/

Kreutz, D., Ramos, F. M., Verissimo, P., Rothenberg, C. E., Azodolmolky, S., & Uhlig, S. (2015). Software-defined networking: A comprehensive survey. *Proceedings of the IEEE, 103*(1), 14–76. doi:10.1109/JPROC.2014.2371999

Krishna, M. B. & Lorenz, P. (2017, December). Delay aware secure hashing for opportunistic message forwarding in Internet of Things. In Proceedings 2017 IEEE Globecom Workshops (GC Wkshps) (pp. 1-6). IEEE. doi:10.1109/GLOCOMW.2017.8269222

Kułakowski, P., Vales-Alonso, J., Egea-López, E., Ludwin, W., & García-Haro, J. (2010). Angle-of-arrival localization based on antenna arrays for wireless sensor networks. *Computers & Electrical Engineering*, *36*(6), 1181–1186. doi:10.1016/j.compeleceng.2010.03.007

Kumar, A., Dhurandher, S. K., Woungang, I., Obaidat, M. S., Gupta, S., & Rodrigues, J. J. (2017). An altruism-based trust-dependent message forwarding protocol for opportunistic networks. *International Journal of Communication Systems*, *30*(10), e3232. doi:10.1002/dac.3232

Kumar, A., Shwe, H. Y., Wong, K. J., & Chong, P. H. (2017). Location-based routing protocols for wireless sensor networks: A survey. *Wireless Sensor Network*, *9*(1), 25–72. doi:10.4236/wsn.2017.91003

Kumari, J., Kumar, P., & Singh, S. K. (2019). Localization in three-dimensional wireless sensor networks: A survey. *The Journal of Supercomputing*, 1–44.

Kumari, S., Khan, M. K., & Atiquzzaman, M. (2015). User authentication schemes for wireless sensor networks: A review. *Ad Hoc Networks*, *27*, 159–194. doi:10.1016/j.adhoc.2014.11.018

Kumar, P., Kunwar, R. S., & Sachan, A. (2016). A survey report on: Security & challenges in internet of things. In *Proc National Conference on ICT & IoT* (pp. 35-39).

Kwon, J. (2018). Tendermint: Consensus without mining (v0.6). Available at https://tendermint.com/static/docs/tendermint.pdf

Larimer, D. (2018). *Transactions as proof-of-stake*. Available at https://bravenewcoin.com/assets/Uploads/TransactionsAsProofOfStake10.pdf

Lazos, L., & Poovendran, R. (2004, October). SeRLoc: Secure range-independent localization for wireless sensor networks. In *Proceedings of the 3rd ACM workshop on Wireless security* (pp. 21-30). ACM. 10.1145/1023646.1023650

Lee, I., & Lee, K. (2015). The Internet of Things (IoT): Applications, investments, and challenges for enterprises. *Business Horizons*, *58*(4), 431–440. doi:10.1016/j.bushor.2015.03.008

Leitao, P., Colombo, A. W., & Karnouskos, S. (2016). Industrial automation based on cyber-physical systems technologies: Prototype implementations and challenges. *Computers in Industry*, *81*, 11–25. doi:10.1016/j.compind.2015.08.004

Lennvall, T., Gidlund, M., & Akerberg, J. (2017). *Challenges when bringing IoT into Industrial Automation*. IEEE Africon. doi:10.1109/AFRCON.2017.8095602

Leung-Yan-Cheong, S., & Hellman, M. (1978). The Gaussian wire-tap channel. *IEEE Transactions on Information Theory*, *24*(4), 451–456. doi:10.1109/TIT.1978.1055917

Li, Z. & Guang, G. (2008). *A survey on security in wireless sensor networks*. Retrieved from HTTP://CACR.UWATERLOO.CA/TECHREPORTS/2008/CACR2008-20.PDF

Liang, L., Zheng, K., Sheng, Q., & Huang, X. (2016). A denial of service attack method for an IoT system. In *Proceedings 8th International Conference on Information Technology in Medicine and Education (ITME)*. IEEE, pp. 360–364. 10.1109/ITME.2016.0087

Li, C., Wang, Z., & Yang, C. (2011). Secure routing for Wireless Mesh Networks. *International Journal of Network Security, 13*(2), 109–120.

Lin, H., Ma, J., Hu, J., & Yang, K. (2012). PA-SHWMP: A privacy aware secure hybrid wireless mesh protocol for IEEE 802.11s wireless mesh networks. *EURASIP Journal on Wireless Communications and Networking, 69*, 1–16. doi:10.1186/1687-1499-2012-69

Li, Q., & Trappe, W. (2006). Light-weight detection of spoofing attacks in wireless networks. In Proceedings *2006 IEEE International Conference on Mobile Ad Hoc and Sensor Systems*, pp. 845–851. 10.1109/MOBHOC.2006.278663

Liu, C. H., Yang, B., & Liu, T. (2014). Efficient naming, addressing and profile services in Internet-of-Things sensory environments. *Ad Hoc Networks, 18*, 85–101. doi:10.1016/j.adhoc.2013.02.008

Liu, C., Ranjan, R., Yang, C., Zhang, X., Wang, L., & Chen, J. (2015). Mur-dpa: Top-down levelled multi-replica merkle hash tree based secure public auditing for dynamic big data storage on cloud. *IEEE Transactions on Computers, 64*(9), 2609–2622. doi:10.1109/TC.2014.2375190

Liu, C., Yang, C., Zhang, X., & Chen, J. (2015). External integrity verification for outsourced big data in cloud and IoT: A big picture. *Future Generation Computer Systems, 49*, 58–67. doi:10.1016/j.future.2014.08.007

Li, X., Li, D., Wan, J., Vasilakos, A. V., Lai, C. F., & Wang, S. (2017). A review of industrial wireless networks in the context of Industry 4.0. *Wireless Networks, 23*(1), 23–41. doi:10.100711276-015-1133-7

Lokhande, D. B., & Dhainje, P. B. (2019). A novel approach for transaction management in heterogeneous distributed real time replicated database systems. *International Journal for Scientific Research and Development, 7*(1), 840–844.

Lopez, J., Rios, R., Bao, F., & Wang, G. (2017). Evolving privacy: From sensors to the Internet of Things. *Future Generation Computer Systems, 75*, 46–57. doi:10.1016/j.future.2017.04.045

Low, K.-S. & Keck, M.-T. (2003). Advanced precision linear stage for industrial automation applications. *IEEE Transactions on Instrumentation and Measurement, 52*(3).

Lundqvist, T. & De Blanche, A. (2017). Thing-to-thing electricity micro payments using blockchain technology. In *Proceedings of the Global Internet of Things Summit (GIoTS)*, Geneva, Switzerland, pp. 6-9.

M. Díaz, C. Martín, B. Rubio. (2016). State-of-the-art, challenges, and open issues in the integration of internet of things and cloud computing. *Journal of Networks Computers*. pp. 99-117.

Makeshwar, P. B., Kalra, A., Rajput, N. S., & Singh, K. P. (2015, February). Computational scalability with Apache Flume and Mahout for large scale round the clock analysis of sensor network data. In *Proceedings 2015 National Conference on Recent Advances in Electronics & Computer Engineering (RAECE)* (pp. 306-311). IEEE. 10.1109/RAECE.2015.7510212

Malik, R., Mittal, M., Batra, I., & Kiran, C. (2011). Wireless Mesh Networks (WMN). *International Journal of Computers and Applications*, *1*(23), 68–76. doi:10.5120/533-697

Maraiya, K., Kant, K., & Gupta, N. (2011). Efficient cluster head selection scheme for data aggregation in wireless sensor network. *International Journal of Computers and Applications*, *23*(9), 10–18. doi:10.5120/2981-3980

Maram, B., Gnanasekar, J. M., Manogaran, G., & Balaanand, M. (2019). Intelligent security algorithm for UNICODE data privacy and security in IOT. *Service Oriented Computing and Applications*, *13*(1), 3–15. doi:10.100711761-018-0249-x

Martignon, F., Paris, S., & Capone, A. (2009). A framework for detecting selfish misbehaviour in Wireless Mesh Community Networks. In *Proceedings of 5th ACM International Symposium on QoS and Security for Wireless and Mobile Networks, Tenerife, Canary Islands, Spain*, 65-72. doi:10.1145/1641944.1641958

Marti, S., Giuli, T., Lai, K., & Baker, M. (2000). Mitigating routing misbehaviour in mobile ad hoc networks. In *Proceedings of Sixth Annual International Conference on Mobile Computing and Networking, Boston*, MA, 255-265. doi:1910.3459550.1145/345

Mathiason, G., Andler, S. F., & Son, S. H. (2007, August). Virtual full replication by adaptive segmentation. *Proceedings of the 13th IEEE International Conference on Embedded and Real-Time Computing Systems and Applications (RTCSA 2007)* (pp. 327-336). IEEE.

Mavinkattimath, S. G., Khanai, R., & Torse, D. A. (2019, April). A survey on secured wireless body sensor networks. In *2019 International Conference on Communication and Signal Processing (ICCSP)* (pp. 0872-0875). IEEE. 10.1109/ICCSP.2019.8698032

Mavromoustakis, C. X., Mastorakis, G., & Batalla, J. M. (Eds.). (2016). *Internet of Things (IoT) in 5G mobile technologies* (Vol. 8). Springer. doi:10.1007/978-3-319-30913-2

Mazieres, D. (2018). *The stellar consensus protocol: A federated model for internet-level consensus.* Available at papers/stellar-consensus-protocol.pdf

McLaughlin, P., & McAdam, R. (2016). *The undiscovered country: The future of industrial automation. Honeywell Process Solutions*. Honeywell.

Meidan, Y., Bohadana, M., Shabtai, A., Ochoa, M., Tippenhauer, N. O., Guarnizo, J. D., & Elovici, Y. (2017a). Detection of unauthorized IoT devices using machine learning techniques. Retrieved from https://arxiv.org/abs/1709.04647

Meidan, Y., Bohadana, M., Shabtai, A., Ochoa, M., Tippenhauer, N. O., Guarnizo, J. D., & Elovici, Y. (2017b). ProfilIoT: A machine learning approach for IoT device identification based on network traffic analysis. *Proceedings of the Symposium on Applied Computing*. 10.1145/3019612.3019878

Michie, D., Spiegelhalter, D. J., & Taylor, C. C. (1994). Machine learning. *Neural and Statistical Classification, 13*.

Middleware. Oxford English Dictionary. Online. Available at https://en.oxforddictionaries.com/definition/middleware

Miettinen, M., Marchal, S., Hafeez, I., Asokan, N., Sadeghi, A. R., & Tarkoma, S. (2017) IoT Sentinel: Automated device-type identification for security enforcement in IoT. *Proceedings of IEEE International Conference on Distributed Computing Systems*, 283.

Mirkovic, J., & Reiher, P. (2004). A taxonomy of DDoS attack and DDoS defense mechanisms. *Computer Communication Review*, *34*(2), 39–53. doi:10.1145/997150.997156

Modum. (2017). Available at https://modum.io/

Mosenia, A. (2018). Addressing security and privacy challenges in internet of things. *arXiv preprint arXiv:1807.06724*.

Moustafa, N., Creech, G., & Slay, J. (2017). Big data analytics for intrusion detection system: Statistical decision-making using finite dirichlet mixture models. In *Data analytics and decision support for cybersecurity* (pp. 127–156). Cham, Switzerland: Springer. doi:10.1007/978-3-319-59439-2_5

Moustafa, N., Hu, J., & Slay, J. (2019). A holistic review of network anomaly detection systems: A comprehensive survey. *Journal of Network and Computer Applications*, *128*, 33–55. doi:10.1016/j.jnca.2018.12.006

Mumtaz, S., Asohaily, A., Pang, Z., Rayes, A., Tsang, K. F., & Rodriguez, J. (2017). Massive internet of things for industrial applications. *IEEE Industrial Electronics Magazine*, 28–33.

Naveed, A., Kanhere, S. S., & Jha, S. K. (2009). Attacks and security mechanisms security in wireless mesh networks. Boca Raton, FL: Auerbach Publications.

Ndiaye, M., Hancke, G., & Abu-Mahfouz, A. (2017). Software defined networking for improved wireless sensor network management: A survey. *Sensors (Basel)*, *17*(5), 1031. doi:10.339017051031 PMID:28471390

Nie, T., & Zhang, T. (2009, January). A study of DES and Blowfish encryption algorithm. *Proceedings of the Tencon 2009-2009 IEEE Region 10 Conference* (pp. 1-4). IEEE.

Nivash, J. P., Raj, E. D., Babu, L. D., Nirmala, M., & Kumar, V. M. (2014, July). Analysis on enhancing storm to efficiently process big data in real time. In *Proceedings Fifth International Conference on Computing, Communications and Networking Technologies (ICCCNT)* (pp. 1-5). IEEE. 10.1109/ICCCNT.2014.7093076

Noel, A. B., Abdaoui, A., Elfouly, T., Ahmed, M. H., Badawy, A., & Shehata, M. S. (2017). Structural health monitoring using wireless sensor networks: A comprehensive survey. *IEEE Communications Surveys and Tutorials, 19*(3), 1403–1423. doi:10.1109/COMST.2017.2691551

Noubir, G., & Lin, G. (2003). Low-power DoS attacks in data wireless LANs and countermeasures, SIGMOBILE Mob. *Computer Communication Review, 7*(3), 29–30. doi:10.1145/961268.961277

Novinson, M. (2018). *8 biggest DDoS attacks today and what you can learn from them.* Retrieved from https://www.crn.com/slide-shows/security/8-biggest-ddos-attacks-today-and-what-you-can-learn-from-them

Noyes, C. (2016a). Bitav: Fast anti-malware by distributed blockchain consensus and feedforward scanning. *arXiv preprint arXiv:1601.01405*

Nunes, B. A. A., Mendonca, M., Nguyen, X. N., Obraczka, K., & Turletti, T. (2014). A survey of software-defined networking: Past, present, and future of programmable networks. *IEEE Communications Surveys and Tutorials, 16*(3), 1617–1634. doi:10.1109/SURV.2014.012214.00180

O' Gorman, A. (2016). *Internet of Things in the Industrial Sector.* IBM Global Business Services, IBM Corporation.

Oguntimilehin, A., & Ademola, E. O. (2014). A review of big data management, benefits and challenges. *A Review of Big Data Management, Benefits and Challenges, 5*(6), 1–7.

Oliviero, F., & Romano, S. P. (2008). A reputation-based metric for secure routing in wireless mesh networks. In *Proceedings of IEEE Global Telecommunications Conference, New Orleans, LA,* 1–5. 10.1109/GLOCOM.2008.ECP.374

OneM2M. (2017). Security solutions–OneM2M technical specification. Retrieved from http://onem2m.org/technical/latest-drafts

Ongaro, D. & Ousterhout, J. (2014, June). In search of an understandable consensus algorithm. In *Proceedings of USENIX Annual Technical Conference,* Philadelphia, PA.

Oussous, A., Benjelloun, F. Z., Lahcen, A. A., & Belfkih, S. (2018). Big Data technologies: A survey. *Journal of King Saud University-Computer and Information Sciences, 30*(4), 431–448. doi:10.1016/j.jksuci.2017.06.001

Outchakoucht, A., Hamza, E. S., & Leroy, J. P. (2017). Dynamic access control policy based on blockchain and machine learning for the internet of things. *International Journal of Advanced Computer Science and Applications*, 8(7), 417–424. doi:10.14569/IJACSA.2017.080757

Ouzzani, M., Medjahed, B., & Elmagarmid, A. K. (2009). Correctness criteria beyond serializability. In *Encyclopedia of database systems* (pp. 501–506). Academic Press.

Owais, S. S., & Hussein, N. S. (2016). Extract five categories CPIVW from the 9V's characteristics of the Big Data. *International Journal of Advanced Computer Science and Applications*, 7(3), 254–258.

Padhi, S., Tiwary, M., Priyadarshini, R., Panigrahi, C. R., & Misra, R. (2016, March). SecOMN: Improved security approach for opportunistic mobile networks using cyber foraging. In *Proceedings 2016 3rd International Conference on Recent Advances in Information Technology (RAIT)* (pp. 415-421). IEEE.

Padmavathi, B., & Kumari, S. R. (2013). *A survey on performance analysis of DES, AES and RSA algorithm along with LSB substitution*. India: IJSR.

Pai, S., Meingast, M., Roosta, T., Bermudez, S., Wicker, S. B., Mulligan, D. K., & Sastry, S. (2008). Transactional confidentiality in sensor networks. *IEEE Security and Privacy*, 6(4), 28–35. doi:10.1109/MSP.2008.107

Pandey, O. J., Kumar, A., & Hegde, R. M. (2016, March). Localization in wireless sensor networks with cognitive small world characteristics. In *2016 Twenty Second National Conference on Communication (NCC)* (pp. 1-6). IEEE. 10.1109/NCC.2016.7561180

Park, N., & Kang, N. (2016). Mutual authentication scheme in secure internet of things technology for comfortable lifestyle. *Sensors (Basel)*, 6(1), 20–20. doi:10.339016010020 PMID:26712759

Patel, S. T., & Mistry, N. H. (2017). A review: Sybil attack detection techniques in WSN. In *Proceedings 4th International Conference on Electronics and Communication Systems (ICECS)*. IEEE, pp. 184–188. 10.1109/ECS.2017.8067865

Paul, A., & Sato, T. (2017). Localization in wireless sensor networks: A survey on algorithms, measurement techniques, applications and challenges. *Journal of Sensor and Actuator Networks*, 6(4), 24. doi:10.3390/jsan6040024

Pecorella, T., Brilli, L., & Muchhi, L. (2016). The role of physical layer security in IoT: A novel perspective. *Information, 7*(3).

Peddi, P., & DiPippo, L. C. (2002). A replication strategy for distributed real-time object-oriented databases. *Proceedings Fifth IEEE International Symposium on Object-Oriented Real-Time Distributed Computing. ISIRC 2002* (pp. 129-136). IEEE. 10.1109/ISORC.2002.1003670

Peercoin official web page. Available at https://peercoin.net

Pei-Breivold, H., & Sandstrom, K. (2015). Internet of things for industrial automation- Challenges and technical solutions. In *Proceedings of IEEE International Conference on Data Science and Data Intensive Systems.* 10.1109/DSDIS.2015.11

Pelusi, L., Passarella, A., & Conti, M. (2006). Opportunistic networking: Data forwarding in disconnected mobile ad hoc networks. *IEEE Communications Magazine, 44*(11), 134–141. doi:10.1109/MCOM.2006.248176

Peng, B., & Li, L. (2015). An improved localization algorithm based on genetic algorithm in wireless sensor networks. *Cognitive Neurodynamics, 9*(2), 249–256. doi:10.100711571-014-9324-y PMID:25852782

Perera, C., Jayaraman, P. P., Zaslavsky, A., Christen, P., & Georgakopoulos, D. (2014, January). Mosden: An internet of things middleware for resource constrained mobile devices. In *2014 47th Hawaii International Conference on System Sciences* (pp. 1053-1062). IEEE.

Perera, C., Liu, C. H., Jayawardena, S., & Chen, M. (2014). A survey on internet of things from industrial market perspective. *IEEE Access: Practical Innovations, Open Solutions, 2,* 1660–1679. doi:10.1109/ACCESS.2015.2389854

Pietzuch, P. R. (2004). *Hermes: A scalable event-based middleware (No. UCAM-CL-TR-590).* University of Cambridge, Computer Laboratory.

Ping, Y., Xinghao, J., Yue, W., & Ning, L. (2008). Distributed intrusion detection for mobile adhoc networks. *Elsevier Journal of System Engineering and Electronics, 19*(4), 851-859. doi: (08)60163-2 doi:10.1016/S1004-4132

Prisco, G. (2016) Slock. it to introduce smart locks linked to smart ethereum contracts, decentralize the sharing economy. Bitcoin Magazine. Nov-2015 [Online]. Available at https://bitcoinmagazine. com/articles/sloc-it-to-introduce-smart-locs-lined-to-smart-ethereum-contractsdecentralize-the-sharing-economy-1446746719.

Pu, C. & Leff, A. (1990). Replica control in distributed systems: An asynchronous approach.

Puliafito, A., Cucinotta, A., Minnolo, A. L., & Zaia, A. (2010). Making the internet of things a reality: The wherex solution. In *The Internet of Things* (pp. 99–108). New York, NY: Springer. doi:10.1007/978-1-4419-1674-7_10

Puthal, D., Malik, N., Mohanty, S. P., Kougianos, E., & Yang, C. (2018). Blockchain as decentralized security framework. *IEEE Consumer Electronics Magazine* March 2018. DPOS description on Bitshares. Available at http://docs.bitshares.org/ bitshares/dpos.html

Rashid, B., & Rehmani, M. H. (2016). Applications of wireless sensor networks for urban areas: A survey. *Journal of Network and Computer Applications, 60*, 192–219. doi:10.1016/j.jnca.2015.09.008

Rashid, M., & Chawla, R. (2013). Securing data storage by extending role-based access control. *International Journal of Cloud Applications and Computing, 3*(4), 28–37. doi:10.4018/ijcac.2013100103

Rathore, H., Badarla, V., Jha, S., & Gupta, A. (2014). Novel approach for security in wireless sensor network using bio-inspirations. *Proceedings of International Conference on Communication Systems and Networking.* 10.1109/COMSNETS.2014.6734875

Ray, P. P. (2018). A survey on Internet of Things architectures. *Journal of King Saud University-Computer and Information Sciences, 30*(3), 291–319. doi:10.1016/j.jksuci.2016.10.003

Razzaque, M. A., Milojevic-Jevric, M., Palade, A., & Clarke, S. (2016). Middleware for internet of things: a survey. *IEEE Internet of Things Journal, 3*(1), 70-95.

Reddi, K. K., & Indira, D. (2013). Different technique to transfer Big Data: Survey. *Int. Journal of Engineering Research and Applications, 3*(6), 708–711.

Reed, M. G., Syverson, P. F., & Goldschlag, D. M. (1998). Anonymous connections and onion routing. *IEEE Journal on Selected Areas in Communications, 16*(4), 482–494. doi:10.1109/49.668972

Ren, L. (2018). Proof of stake velocity: Building the social currency of the digital age. Available at https://www.reddcoin.com/papers/PoSV. pdf

Ren, K., Yu, S., Lou, W., & Zhang, Y. (2010). PEACE: A novel privacy-enhanced yet accountable security framework for metropolitan wireless mesh networks. *IEEE Transactions on Parallel and Distributed Systems, 21*(2), 203–215. doi:10.1109/TPDS.2009.59

Reyna, A., Martín, C., Chen, J., Soler, E., & Díaz, M. (2018). On blockchain and its integration with IoT, Challenges and opportunities, *Future Generation Computer Systems*, 88, pp. 173-190.

Riaz, R., Kim, K.-H., & Ahmed, H. F. (2009). Security analysis survey and framework design for IP connected LoWPANs. In *Proceedings 2009 International Symposium on Autonomous Decentralized Systems*, pp. 1–6. 10.1109/ISADS.2009.5207373

Riggins, F. J. & Wamba, S. F. (2015, January). Research directions on the adoption, usage, and impact of the internet of things through the use of big data analytics. In *2015 48th Hawaii International Conference on System Sciences (HICSS)*, (pp. 1531-1540). IEEE. 10.1109/HICSS.2015.186

Román-Castro, R., López, J., & Gritzalis, S. (2018). Evolution and trends in IoT security. *Computer, 51*(7), 16–25. doi:10.1109/MC.2018.3011051

Roman, R., Lopez, J., & Mambo, M. (2018). A survey and analysis of security threats and challenges. *Future Generation Computer Systems*, *78*, 680–698. doi:10.1016/j.future.2016.11.009

Roman, R., Zhou, J., & Lopez, J. (2013). On the features and challenges of security and privacy in distributed internet of things. *Computer Networks*, *57*(10), 2266–2279. doi:10.1016/j.comnet.2012.12.018

Rong, P., & Sichitiu, M. L. (2006, September). *Angle of arrival localization for wireless sensor networks. In 2006 3rd annual IEEE communications society on sensor and ad hoc communications and networks* (Vol. 1, pp. 374–382). Piscataway, NJ: IEEE.

Sadeghi, A.-R., Wachsmann, C., & Waidner, M. (2015). *Security and Privacy Challenges in Industrial Internet of Things, DAC*. San Francisco, CA: ACM.

Salehi, M., Darehshoorzadeh, A., & Boukerche, A. (2015). On the effect of black-hole attack on opportunistic routing protocols. In *12th ACM Symposium on Performance Evaluation of Wireless Ad Hoc, Sensor, & Ubiquitous Networks*. ACM, pp. 93–100.

Salem, N. B., & Hubaux, J. P. (2006). Securing Wireless Mesh Networks. *IEEE Wireless Communications*, *13*(2), 50–55. doi:10.1109/MWC.2006.1632480

Salem, R., Saleh, S. A., & Abdul-Kader, H. (2016). Scalable data-oriented replication with flexible consistency in real-time data systems. *Data Science Journal*, 15.

Salman, T. & Jain, R. (2017). A survey of protocols and standards for internet of things. *Advanced Computing and Communications, 1*(1).

saponlinetutorials.com. (n.d.). What is SAP CRM Middleware. Retrieved from https://www.saponlinetutorials.com/sap-crm-middleware/

Schwartz, D., Youngs, N., & Britto, A. (2014). The ripple protocol consensus algorithm. *White paper, Ripple Labs.*

Sethi, M., Arkko, J., & Kernen, A. (2012). End-to-end security for sleepy smart object networks. In *Proceedings 37th Annual IEEE Conference on Local Computer Networks Workshops*, pp. 964–972.10.1109/LCNW.2012.6424089

Seyedzadegan, M., Othman, M., Ali, B. M., & Subramaniam, S. (2011). Wireless Mesh Networks: WMN Overview, WMN Architecture, *International Conference on Communication Engineering and Networks*, Singapore, 12-18.

Sgora, A., Vergados, D. D., & Chatzimisios, P. (2016). A survey on security and privacy issues wireless mesh networks, security and communication networks. *Wiley Online Library*, *9*(13), 1877–1889. doi:10.1002ec.846

Shanker, U., Misra, M., & Sarje, A. K. (2008). Distributed real time database systems: Background and literature review. *Distributed and Parallel Databases, 23*(2), 127–149. doi:10.100710619-008-7024-5

Shannon, C. E. (1949). Communication theory of secrecy systems. *Bell Labs Technical Journal, 28*(4), 656–715. doi:10.1002/j.1538-7305.1949.tb00928.x

Shen, C. C., Srisathapornphat, C., & Jaikaeo, C. (2001). Sensor information networking architecture and applications. *IEEE Personal Communications, 8*(4), 52-59.

Shila, D. M., & Anjali, T. (2008). Defending selective forwarding attacks in WMNs. In *Proceedings of IEEE International Conference on Electro/Information Technology, USA*, 96-101. doi:10.1109/eit.2008.4554274

Shrivastava, P. & Shanker, U. (2018, August). Replica update technique in RDRTDBS: Issues & challenges. *Proceedings of the 24th International Conference on Advanced Computing and Communications (ADCOM-2018)*, Bangalore, India (pp. 21-23). Academic Press.

Shrivastava, P., Jain, R., & Raghuwanshi, K. S. (2014, January). A modified approach of key manipulation in cryptography using 2d graphics image. *Proceedings of the 2014 International Conference on Electronic Systems, Signal Processing and Computing Technologies* (pp. 194-197). IEEE. 10.1109/ICESC.2014.40

Shrivastava, P., & Shanker, U. (2018). Replica control following 1SR in DRTDBS through best case of transaction execution. In *Advances in Data and Information Sciences* (pp. 139–150). Singapore: Springer. doi:10.1007/978-981-10-8360-0_13

Shrivastava, P., & Shanker, U. (2018). Replication protocol based on dynamic versioning of data object for replicated DRTDBS. *International Journal of Computational Intelligence & IoT, 1*(2).

Shrivastava, P., & Shanker, U. (2019, January). Real time transaction management in replicated DRTDBS. *Proceedings of the Australasian Database Conference* (pp. 91-103). Springer. 10.1007/978-3-030-12079-5_7

Shrivastava, P., & Shanker, U. (2019, January). Supporting transaction predictability in replicated DRTDBS. *Proceedings of the International Conference on Distributed Computing and Internet Technology* (pp. 125-140). Springer. 10.1007/978-3-030-05366-6_10

Siddiqui, M. S., & Hong, C. S. (2007). Security issues in Wireless Mesh Networks. In *Proceedings of International Conference on Multimedia and Ubiquitous Engineering, Seoul, Korea*, 717–722, doi:10.1109/mue.2007.187

Siemens. (2017). *MindSphere The Cloud-based, open IoT operating system for digital transformation.* Siemens PLM Software

Singh, P., & Rashid, E. (2015). Smart home automation deployment on third party cloud using internet of things. *Journal of Bioinformatics and Intelligent Control, 4*(1), 31–34. doi:10.1166/jbic.2015.1113

Sivaraman, V., Gharakheili, H. H., Fernandes, C., Clark, N., & Karliychuk, T. (2018). Smart IoT devices in the home: Security and privacy implications. *IEEE Technology and Society Magazine, 37*(2), 71–79. doi:10.1109/MTS.2018.2826079

Smith, D. B., Hanlen, L. W., Zhang, J. A., Miniutti, D., Rodda, D., & Gilbert, B. (2011). First- and second-order statistical characterizations of the dynamic body area propagation channel of various bandwidths. *Annals of Telecommunications, 66*(3-4), pp. 187-203.

Smith, D., Hanlen, L., Miniutti, D., Zhang, J., Rodda, D., & Gilbert, B. (2008, October). Statistical characterization of the dynamic narrowband body area channel. In *First International Symposium on Applied Sciences on Biomedical and Communication Technologies. ISABEL'08.* (pp. 1-5). IEEE.

Smith, D. B., & Hanlen, L. W. (2015). Channel modeling for wireless body area networks. In *Ultra-Low-Power Short-Range Radios* (pp. 25–55). Cham, Switzerland: Springer. doi:10.1007/978-3-319-14714-7_2

Smith, D. B., Miniutti, D., Lamahewa, T. A., & Hanlen, L. W. (2013). Propagation models for body-area networks: A survey and new outlook. *IEEE Antennas & Propagation Magazine, 55*(5), 97–117. doi:10.1109/MAP.2013.6735479

Sniderman, B., Mahto, M., & Cotteleer, M. J. (2016). *Industry 4.0 and manufacturing ecosystems Exploring the world of connected enterprises*. Deloitte, USA: Deloitte LLP Consulting.

Son, S. H. & Zhang, F. (1995, April). Real-time replication control for distributed database systems: Algorithms and their performance. In DASFAA, 11, pp. 214-221.

Son, S. H., & Kouloumbis, S. (1993). A token-based synchronization scheme for distributed real-time databases. *Information Systems, 18*(6), 375–389. doi:10.1016/0306-4379(93)90014-R

Son, S. H., Zhang, F., & Hwang, B. (1996). Concurrency control for replicated data in distributed real-time systems. *Journal of Database Management, 7*(2), 12–23. doi:10.4018/jdm.1996040102

Srivastava, A., Shankar, U., & Tiwari, S. K. (2012). A protocol for concurrency control in real-time replicated databases system. *International Journal of Computer Networks and Wireless Communications, 2*(3).

Ssu, K. F., Ou, C. H., & Jiau, H. C. (2005). Localization with mobile anchor points in wireless sensor networks. *IEEE Transactions on Vehicular Technology, 54*(3), 1187–1197. doi:10.1109/TVT.2005.844642

Stajano, F., & Anderson, R. (2002). Resurrecting duckling: Security issues for ubiquitous computing. *Supplement to Computer, 35*(4), 22–26. doi:10.1109/mc.2002.1012427

Stallings, W. (2006). Network security essentials (3rd ed.). Upper Saddle River, NJ: Prentice Hall.

Suciu, G., Suciu, V., Martian, A., Craciunescu, R., Vulpe, A., Marcu, I., & Fratu, O. (2015). Big data, internet of things and cloud convergence–an architecture for secure e-health applications. *Journal of Medical Systems, 39*(11), 141. doi:10.100710916-015-0327-y PMID:26345453

Syberfeldt, S. (2007). *Optimistic replication with forward conflict resolution in distributed real-time databases* [Doctoral dissertation]. Institutionen för datavetenskap.

Tang, C., & Wu, D. O. (2008). An efficient mobile authentication scheme for wireless networks. *IEEE Transactions on Wireless Communications, 7*(4), 1408–1416. doi:10.1109/TWC.2008.061080

Terziyan, V., Kaykova, O., & Zhovtobryukh, D. (2010, May). Ubiroad: Semantic middleware for context-aware smart road environments. In *2010 Fifth International Conference on Internet and Web Applications and Services* (pp. 295-302). IEEE. 10.1109/ICIW.2010.50

Thi, T. H. L., Palopoli, L., Passerone, R., Ramadian, Y., & Cimatti, A. (2010). *Parametric analysis of distributed firm real-time systems: A case study.* Piscataway, NJ: IEEE.

Tiago, M. F.-C. & Fraga-Lamas, P. (2018). A review on the use of blockchain for the Internet of Things. *IEEE Transactions.*

Tian, F. (2016, June). An agri-food supply chain traceability system for china based on RFID & blockchain technology. In *Proceedings of the 13th International Conference on Service Systems and Services Management*, Kunming, China.

Tomic, S., Beko, M., & Dinis, R. (2016). Distributed RSS-AoA based localization with unknown transmit powers. *IEEE Wireless Communications Letters, 5*(4), 392–395. doi:10.1109/LWC.2016.2567394

Tong, Y. P., Liu, N., & Wu, Y. (2009). Security in wireless mesh networks: Challenges and solutions. In *Proceedings of Sixth International Conference on Information Technology: New Generations*, Las Vegas, NV, 423-428. doi:10.1109/ITNG.2009.20

Tönjes, R., Barnaghi, P., Ali, M., Mileo, A., Hauswirth, M., Ganz, F., & Puiu, D. (2014, June). Real time IoT stream processing and large-scale data analytics for smart city applications. In *poster session, European Conference on Networks and Communications.* sn.

Tourrilhes, J., Sharma, P., Banerjee, S., & Pettit, J. (2014). SDN and OpenFlow Evolution: A standards perspective. *Computer, 47*(11), 22–29. doi:10.1109/MC.2014.326

Trong, M. H., Dinh, V. L., & Kim, N. Q. (2013). A study on routing performance of 802.11 based wireless mesh networks under serious attacks. In *Proceedings of International Conference on Computing, Management and Telecommunications*, 295-297. doi:10.1109/ComManTel.2013.6482408

Tsai, C. F., Hsu, Y. F., Lin, C. Y., & Lin, W. Y. (2009). Intrusion detection by machine learning: A review. *Expert Systems with Applications*, *36*(10), 11994–12000. doi:10.1016/j.eswa.2009.05.029

Tuna, G., & Gungor, V. C. (2017). A survey on deployment techniques, localization algorithms, and research challenges for underwater acoustic sensor networks. *International Journal of Communication Systems*, *30*(17), e3350. doi:10.1002/dac.3350

Ulusoy, Ö. (1994). Processing real-time transactions in a replicated database system. *Distributed and Parallel Databases*, *2*(4), 405–436. doi:10.1007/BF01265321

Vaghefi, R. M., & Buehrer, R. M. (2015). Cooperative joint synchronization and localization in wireless sensor networks. *IEEE Transactions on Signal Processing*, *63*(14), 3615–3627. doi:10.1109/TSP.2015.2430842

Vashishth, V., Chhabra, A., & Sharma, D. K. (2019). GMMR: A Gaussian mixture model based unsupervised machine learning approach for optimal routing in opportunistic IoT networks. *Computer Communications*, *134*, 138–148. doi:10.1016/j.comcom.2018.12.001

Veena, P., Panikkar, S., Nair, S., & Brody, P. (2015). Empowering the edge-practical insights on a decentralized internet of things. *IBM Institute for Business*, 17.

Vetriselvi, V., Shruti, P. S., & Abraham, S. (2018, January). Two-level intrusion detection system in SDN using machine learning. In *Proceedings International Conference on Communications and Cyber Physical Engineering 2018* (pp. 449-461). Springer, Singapore.

Vijay, U. & Gupta, N. (2013, January). Clustering in WSN based on minimum spanning tree using divide and conquer approach. In Proceedings of World Academy of Science, Engineering and Technology 79, p. 578. World Academy of Science, Engineering and Technology (WASET).

Vural, S., Wei, D., & Moessner, K. (2013). Survey of experimental evaluation studies for wireless mesh network deployments in urban areas towards ubiquitous internet. *IEEE Communications Surveys and Tutorials*, *15*(1), 223–239. doi:10.1109/SURV.2012.021312.00018

Wan, J., Tang, S., Shu, Z., Di Li, S. W., Imran, M., & Vasilakos, A. V. (2016). Software-defined industrial internet of things in the context of industry 4.0, IEEE Sensors Journal, 16(20).

Wang, C., Wang, Q., Ren, K., & Lou, W. (2010). Privacy-preserving public auditing for data storage security in cloud computing. In *Proceedings INFOCOM, San Diego*, CA, IEEE, 2010, pp. 1–9. 10.1109/INFCOM.2010.5462173

Wang, C., & Zhang, Y. (2015). New authentication scheme for wireless body area networks using the bilinear pairing. *Journal of Medical Systems*, *39*(11), 136. doi:10.100710916-015-0331-2 PMID:26324170

Wang, F., Yao, L. W., & Yang, Y. L. (2011). Efficient verification of distributed real-time systems with broadcasting behaviors. *Real-Time Systems, 47*(4), 285–318. doi:10.100711241-011-9122-0

Wang, W., Kong, J., Bhargava, B., & Gerla, M. (2008). Visualisation of wormholes in underwater sensor networks: A distributed approach. *Int. J. Secur. Netw., 3*(1), 10–23. doi:10.1504/IJSN.2008.016198

Wang, X., Ning, Z., Zhou, M., Hu, X., Wang, L., Hu, B., ... Guo, Y. (2018). A privacy-preserving message forwarding framework for opportunistic cloud of things. *IEEE Internet of Things Journal, 5*(6), 5281–5295. doi:10.1109/JIOT.2018.2864782

Wang, X., Patil, A., & Wang, W. (2006). VoIP over wireless mesh networks Challenges and approaches. In *Proceedings of 2nd Annual International Workshop on Wireless Internet*, 1-9. doi:10.1145/1234161.1234167

Wang, X., Wong, J. S., Stanley, F., & Basu, S. (2009). Cross-layer based anomaly detection in Wireless Mesh Networks. In *Proceedings of 9th Annual International Symposium on Applications and the Internet*, Bellevue, WA, 9–15. 10.1109/SAINT.2009.11

Wang, X., Wong, J., & Zhang, W. (2008). A heterogeneity-aware framework for group key-management in wireless mesh networks. In *Proceedings of 4th International Conference on Security and Privacy in Communication Networks*, Instabul, Turkey. 10.1145/1460877.1460918

Weekly, K. & Pister, K. (2012). Evaluating sinkhole defense techniques in RPL networks. In *Proceedings of the 2012 20th IEEE International Conference on Network Protocols (ICNP)*, IEEE Computer Society, Washington, DC. . pp 1-6.10.1109/ICNP.2012.6459948

Wikipedia. Internet of Things. Retrieved from https://en.wikipedia.org/wiki/Internet_of_things

Wollschlaeger, M., Sauter, T., & Jasperneite, J. (2017). *The future of industrial communication automation networks in the era of the internet of things and industry 4.0. IEEE Industrial Electronics Magazine.*

Wong, K. H. M., Zheng, Y., Cao, J., & Wang, S. (2006). A dynamic user authentication scheme for wireless sensor networks. In *Proceedings of IEEE International Conference Sensor Networks, Ubiquitous, Trustworthy Computing*, 244–251. IEEE Computer Society. 10.1109/SUTC.2006.1636182

Wu, L., Zhang, Y., Li, L., & Shen, J. (2016). Efficient and anonymous authentication scheme for wireless body area networks. *Journal of Medical Systems, 40*(6), 134. doi:10.100710916-016-0491-8 PMID:27091755

Wu, X., & Li, N. (n.d.). Achieving privacy in mesh networks. *4th ACM Workshop on Security of Ad Hoc and Sensor Networks*, Alexandria, VA, 13–22. doi:10.1145/1180345.1180348

Wyner, A. D. (1975). The wire-tap channel. *Bell Labs Technical Journal, 54*(8), 1355–1387. doi:10.1002/j.1538-7305.1975.tb02040.x

Xiao, L., Wan, X., Dai, C., Du, X., Chen, X., & Guizani, M. (2018). Security in mobile edge caching with reinforcement learning.

Xiao, J., Zhou, Z., Yi, Y., & Ni, L. M. (2016). A survey on wireless indoor localization from the device perspective. [CSUR]. *ACM Computing Surveys, 49*(2), 25. doi:10.1145/2933232

Xiao, L., Greenstein, L. J., Mandayam, N. B., & Trappe, W. (2009). Channel-based detection of sybil attacks in wireless networks. *IEEE Transactions on Information Forensics and Security, 4*(3), 492–503. doi:10.1109/TIFS.2009.2026454

Xiong, M., Ramamritham, K., Haritsa, J. R., & Stankovic, J. A. (2002). MIRROR: A state-conscious concurrency control protocol for replicated real-time databases. *Information Systems, 27*(4), 277–297. doi:10.1016/S0306-4379(01)00053-9

Xu, W., Trappe, W., Zhang, Y., & Wood, T. (2005). The feasibility of launching and detecting jamming attacks in wireless networks. In *Proceedings of the 6th ACM International Symposium on Mobile Ad Hoc Networking and Computing, nMobiHoc '05*, ACM, New York, NY. . pp. 46–57.10.1145/1062689.1062697

*Yamaji, M., Ishii, Y., Shimamura, T., & Yamamoto, S. (n.d.). Wireless sensor network for industrial automation.* Tokyo, Japan: Ubiquitous Field Computing Research Centre, Yokogawa Electric Corporation.

Yang, Y., Zeng, P., Yang, X., & Huang, Y. (2010). Efficient intrusion detection system model in Wireless Mesh Network; Networks Security Wireless. In *Proceedings of 2nd International Conference on Communications and Trusted Computing*, Wuhan, China, 393–395. doi:10.1109/NSWCTC.2010.226

Yan, J., & Jin, D. (2015, June). VT-Mininet: Virtual-time-enabled mininet for scalable and accurate software-defined network emulation. In *Proceedings of the 1st ACM SIGCOMM Symposium on Software Defined Networking Research* (p. 27). ACM.

Yassin, A., Nasser, Y., Awad, M., Al-Dubai, A., Liu, R., Yuen, C., & Aboutanios, E. (2016). Recent advances in indoor localization: A survey on theoretical approaches and applications. *IEEE Communications Surveys and Tutorials, 19*(2), 1327–1346. doi:10.1109/COMST.2016.2632427

Ye, J., Cheng, X., Zhu, J., Feng, L., & Song, L. (2018). A DDoS attack detection method based on SVM in software defined network. *Security and Communication Networks*.

Yi, D., Xu, G., & Minqing, Z. (2011). The research on certificate less hierarchical key management in Wireless Mesh Network. In *Proceedings of 3rd IEEE International Conference Communication Software and Networks*, Xian, China, 504-507. doi:10.1109/iccsn.2011.6013643

Yi, P., Dai, Z., Zhang, S., Zhong, Y., & ... . (2005). A new routing attack in mobile ad hoc networks. *International Journal of Information Technology*, *11*(2), 83–94.

Young, M., & Boutaba, R. (2011). Overcoming adversaries in sensor networks: A survey of theoretical models and algorithmic approaches for tolerating malicious interference. *IEEE Communications Surveys and Tutorials*, *13*(4), 617–641. doi:10.1109/SURV.2011.041311.00156

Yu, Y., Guo, L., Liu, Y., Zheng, J., & Zong, Y. (2018). An efficient SDN-based DDoS attack detection and rapid response platform in vehicular networks. *IEEE Access: Practical Innovations, Open Solutions*, *6*, 44570–44579. doi:10.1109/ACCESS.2018.2854567

Zaidi, Z. R., Hakami, S., Landfeldt, B., & Moors, T. (2009). Detection and identification of anomalies in wireless mesh networks using Principal Component Analysis (PCA). *Journal of Interconnection Networks*, *10*(04), 517–534. doi:10.1142/S0219265909002698

Zhang, Y., Lee, W., & Huang, Y. (2003). Intrusion detection techniques for mobile wireless networks, *ACM/Kluwer Wireless Networks Journal*, *9*(5), 1-16.

Zhang, Y., Luo, J., & Hu, H. (2006). Wireless mesh networking: Architectures, protocols and standards. Boca Raton, FL: Taylor & Francis Group.

Zhang, H., Zhao, W., Moser, L. E., & Melliar-Smith, P. M. (2011). Design and implementation of a byzantine fault tolerance framework for non-deterministic applications. *IET Software*, *5*(3), 342–356. doi:10.1049/iet-sen.2010.0013

Zhang, L., Song, J., & Pan, J. (2016). A privacy-preserving and secure framework for opportunistic routing in DTNs. *IEEE Transactions on Vehicular Technology*, *65*(9), 7684–7697. doi:10.1109/TVT.2015.2480761

Zhangm, Z., Nait-Abdesselam, F., Ho, P.-H., & Lin, X. (2008). *RADAR: A reputation-based scheme for detecting anomalous nodes in wireless mesh networks.* In Proceedings of *IEEE Wireless Communications & Networking Conference*, Las Vegas, NV. (pp. 2621–2626). doi:10.1109/wcnc.2008.460

Zhang, Y., & Fang, Y. (2006). ARSA: An attack resilient security architecture for multi-hop wireless mesh network. *IEEE Journal on Selected Areas in Communications*, *24*(10), 1916–1928. doi:10.1109/JSAC.2006.877223

Zhao, W. (2014). *Building dependable distributed systems*. John Wiley & Sons. doi:10.1002/9781118912744

Zhen, B. (2008). Body area network (BAN) technical requirements. *15-08-0037-03-0006-ieee-802-15-6-technical-requirements-document-v-5-0. doc*.

Zheng, Z., Xie, S., Dai, H.-N., Chen, X., & Wang, H. (2018). Blockchain challenges and opportunities: A survey. *International Journal of Web and Grid Services*, *14*(4), 352. doi:10.1504/IJWGS.2018.095647

Zhou, X., & McKay, M. R. (2010). Secure transmission with artificial noise over fading channels: Achievable rate and optimal power allocation. *IEEE Transactions on Vehicular Technology*, *59*(8), 3831–3842. doi:10.1109/TVT.2010.2059057

Zhu, G., & Hu, J. (2014). A distributed continuous-time algorithm for network localization using angle-of-arrival information. *Automatica*, *50*(1), 53–63. doi:10.1016/j.automatica.2013.09.033

Zou, Y., & Wang, G. (2016). Intercept behavior analysis of industrial wireless sensor networks in the presence of eavesdropping attack. *IEEE Transactions on Industrial Informatics*, *12*(2), 780–787. doi:10.1109/TII.2015.2399691

Zou, Y., Wang, X., & Shen, W. (2013). Optimal relay selection for physical-layer security in cooperative wireless networks. *IEEE Journal on Selected Areas in Communications*, *31*(10), 2099–2111. doi:10.1109/JSAC.2013.131011

Zou, Y., Wang, X., & Shen, W. (2013). Physical-layer security with multiuser scheduling in cognitive radio networks. *IEEE Transactions on Communications*, *61*(12), 5103–5113. doi:10.1109/TCOMM.2013.111213.130235

Zou, Y., Wang, X., Shen, W., & Hanzo, L. (2014). Security versus reliability analysis of opportunistic relaying. *IEEE Transactions on Vehicular Technology*, *63*(6), 2653–2661. doi:10.1109/TVT.2013.2292903

Zou, Y., Zhu, J., Wang, X., & Leung, V. C. (2015). Improving physical-layer security in wireless communications using diversity techniques. *IEEE Network*, *29*(1), 42–48. doi:10.1109/MNET.2015.7018202

Zou, Y., Zhu, J., Zheng, B., & Yao, Y. D. (2010). An adaptive cooperation diversity scheme with best-relay selection in cognitive radio networks. *IEEE Transactions on Signal Processing*, *58*(10), 5438–5445. doi:10.1109/TSP.2010.2053708

# About the Contributors

**Priyanka Ahlawat** is working as an Assistant Professor in the Department of Computer Engineering, NIT, Kurukshetra, Haryana-India. Dr. Ahlawat obtained her Ph.D. (Computer Engineering) and M. Tech (Computer Engineering) from National Institute of Technology (Institute of National Importance) Kurukshetra (Haryana) and Guru Jambheshwar University of Science and Technology, Hisar, respectively. She has more than 13 years of experience in teaching and research. She has published approximately 25 research papers in various international journals and conferences. She has coordinated several training programs for students and faculty. She has delivered number of expert lectures on different topics. Her research interests include IoT security, information security, wireless sensor networks and cyber security.

**Mayank Dave** received BSc (Engg.) degree from Aligarh Muslim University, Aligarh, India in 1989, M.Tech. degree in Computer Science and Technology and the PhD degree, both from IIT Roorkee, India in 1991 and 2002, respectively. He is a Professor in the Department of Computer Engineering at National Institute of Technology Kurukshetra (NIT Kurukshetra), India with more than 28 years' experience of academic and administrative affairs in the institute. He has published approximately 170 research papers in various international/national journals and conferences. He is a reviewer of several reputed research journals. He has coordinated several research and development projects in the department and institute. He has written a text book titled "Computer Networks" published by Cengage Learning, India. Prof Mayank Dave has guided fifteen PhDs and several M.Tech. and B.Tech. thesis and projects. He is currently guiding several PhDs. His research interests include mobile networks, cyber security, cloud computing, web technologies, etc. He is a Senior Member of the IEEE and the IEEE Computer Society, and member of the ACM, IETE, Computer Society of India, and Institution of Engineers (India).

\* \* \*

**Nazir Ahmad** is currently working as an Assistant Professor in Department of Information Systems, Community College, King Khalid University, Kingdom of Saudi Arabia. The main area of interest is Cloud Computing, Internet of Things and Machine Learning. The author has published several research papers in cloud computing, Internet of Things and machine learning indexed in Scopus and Web of Science-based journals.

**Bhanu Chander** is a research scholar at Pondicherry University, India. He is presently working on wireless sensor networks and machine learning techniques. Graduated from Acharya Nagarjuna University in 2013 and post graduated from the Central University of Rajasthan. Machine learning, IoT, wireless sensor networks, deep learning are his research areas.

**Sanjay Kumar Dhurandher** received the M. Tech. and Ph.D. Degrees in Computer Sciences from the Jawaharlal Nehru University, New Delhi, India. He is presently working as Professor and Head in the Department of Information Technology, Netaji Subhas University of Technology (formerly NSIT), University of Delhi, India. Prior to this, from 1995 to 2000 he worked as a Scientist/Engineer at the Institute for Plasma Research, Gujarat, India which is under the Department of Atomic Energy, India. His current research interests include wireless ad-hoc networks, sensor networks, computer networks, opportunistic networks, network security and Underwater Sensor Networks, IoT. He is serving as the Associate Editor of Wiley's International Journal of Communication Systems and Senior IEEE member.

**Vishal Goyal** is presently working as a Professor in Department of Computer Science, Punjabi University, Patiala, India. He is Coordinator, Research Center for Technology Development for Differently Abled people and co-coordinator of Center for Artificial Intelligence and Data Science, Punjabi University, Patiala. His main research area is cognitive computing, artificial intelligence, natural language processing and machine translation language technologies. He was awarded the Young Scientist Award in 2005. He has copyrighted software Hindi to Punjabi Machine Translation System and the software automatic translation of English to ISL Synthetic Videos. He has worked as the Principal Investigator for Indian Languages Corpora Initiative (ILCI) funded under the Technology Development for Indian Languages (TDIL) program by Ministry of Communications and Information Technology. He is currently working on research funded project of Plagiarism Detection Tool Development for Indian Languages with Special Focus on Hindi and Punjabi, funded by R&D in Electronics Group (Innovation and IPR Division), Ministry of Electronics and Information Technology, Govt. of India. He has published papers in indexed journals and guided number of research scholars at Doctorate level.

**Nitin Gupta** is serving as an Assistant Professor with the Department of Computer Science and Engineering, NIT Hamirpur, Himachal Pradesh, India from 2007. Currently, he is also working toward his Ph.D. at Netaji Subhas Institute of Technology, University of Delhi, New Delhi, India under Prof. Sanjay K. Dhurandher. His research interest includes next generation wireless networks, in particular, cognitive radio networks. He has published various research papers in international journals and conferences of repute like IEEE ICC and IEEE GLOBECOM. He is a senior member of the IEEE Communication Society and the IEEE Technical Committee on Cognitive Networks and a member of the ACM as well. Along with organizing various short-term courses and conferences, he was also a member of technical program committees of various IEEE/ACM/Springer/Scopus conferences. He is a reviewer of the IEEE Systems Journal.

**Sachin Kumar Gupta** received his B.Tech in Electronics and Telecommunication Engineering from National Institute of Technology (NIT), Raipur, Chhattisgarh, India in 2008 and M.Tech & Ph.D. with specialization in Systems Engineering (Wireless Communication), Department of Electrical Engineering from Indian Institute of Technology (Banaras Hindu University) (IIT (BHU)), Varanasi, Uttar Pradesh, India in 2011 & 2016 respectively. Previously, he was the former research fellow in Mobile Computing and Broadband Networking Lab (MBL), Department of Computer Science at National Chiao Tung University (NCTU), Hsinchu, Taiwan (Republic of China), and IIT Jodhpur, India. Currently, he is working as an Assistant Professor in the Department of Electronics and Communication Engineering, Shri Mata Vaishno Devi University, Kakryal, Katra, (J. & K.) since January 1st, 2015. His research interest includes wireless networking, cryptography & network security, software defined networking, and free space optical communication. He has published many papers and book chapters in the international/national journals and prestigious conference proceedings in India as well as abroad. He has served as a coordinator, organizing committee members, session chair of the various workshop, seminar, and conferences. He is also serving as Central Coordinator of Spoken Tutorial Project, IIT Bombay, (GoI, MHRD). He has guided a number of B. Tech projects and M.Tech theses.

**Samyak Jain** is currently pursuing a B.E. (Hons.) Computer Science and Engineering, National Institute of Technology Karnataka, India. He is an avid reader and a research scholar and has an impeccable academic record. He is deeply interested in business transformation using digital technologies. He has been researching, writing and debating on various topics like automation, data analytics, and insights. He has several papers spanning machine learning, Internet of Things and computer networks that are under publication. He is an active member of professional societies like the IEEE (student chapter) and CSI. He has coordinated and led multiple project teams in the institute to create innovative solutions for tough business problems. His areas of interest in research include: data sciences – machine learning and artificial intelligence, computer networks, internet of things, operating systems and knowledge management.

**Deepti Kakkar**, born in 1982, in Jalandhar, Punjab, India. She did her Bachelor of Technology in Electronics and Communication Engineering from Himachal Pradesh University, India in 2003 and Master of Engineering in electronics product design and technology from Punjab University, Chandigarh. She did her PhD from Dr. B.R. Ambedkar National Institute of Technology, Jalandhar, India. She has a total academic experience of 14 years and at present she is an Assistant Professor in Electronics and Communication department with Dr. B. R. Ambedkar National Institute of Technology, Jalandhar, India. Earlier, she has worked as a lecturer in the Electronics and Communication department with the DAV Institute of Engineering and Technology, Jalandhar, Punjab. She has guided more than 15 post graduate engineering dissertations and currently supervising 2 PhD candidates. Her recent research interests include wireless communications, wireless sensor network, neuro developmental disorders, dynamic spectrum allocation, spectrum sensing, software defined radios and cognitive radios.

**Parveen Kakkar** is working as an Assistant Professor (CSE), DAVIET, Jalandhar &amp; pursuing his PhD from IKGPTU, Kapurthala. He has done B.Tech (CSE) &amp; M. Tech (CSE) with Hons. from IKGPTU, Kapurthala. Mr. Parveen Kakkar authored and co-authored several national and international publications. His general research interests are in the areas of network security and computer networks. specific research interests include information security, IDS, network architecture/protocols, network modeling. He has guided 24 M.Tech dissertations and has experience of 19 years in teaching and research.

**Nisha Kandhoul** is a Research Scholar at NSIT (University of Delhi), New Delhi, India. She is presently working on Security Enhancement Techniques for Secure Routing in OppIoT. She graduated from DCRUST, Murthal, Haryana in 2012 and post graduated from NSIT in 2014. She served as Data Developer at Dunnhumby, India for a year. Her area of interest are Security and Privacy techniques, Machine Learning, Cryptography.

**Chandrasekaran Kandasamy** is currently a Professor in the Department of Computer Science &amp; Engineering, National Institute of Technology Karnataka, India, having 31 years of experience. He has more than 300 research papers published by various reputed and peer-reviewed international journals, and conferences. He has received best paper awards and best teacher awards. He serves as a member of various reputed professional societies including the IEEE (Senior Member), the ACM (Senior Member), the CSI (Life Member), ISTE (Life Member) and the Association of British Scholars (ABS). He is also a member in IEEE Computer Society's Cloud Computing STC. He is in the Editorial Team of IEEE Transactions on Cloud Computing, one of the recent and reputed journals of IEEE publication. He has coordinated many sponsored projects, and, some consultancy projects. He has organized numerous events such as International conferences, International Symposium, workshops and several academic short-term programs at NITK. He was a visiting fellow at LMU Leeds, UK in 1995, Visiting Professor at AIT, Bangkok in 2007, and Visitor at UF, USA in 2008 and a Visitor at Univ. of Melbourne, CLOUDS LAB in 2012. He had also worked as Visiting (Professor) at DoMS, IIT Madras during Feb-Dec. 2010. His areas of interest - research include: computer networks and distributed computing, federated cloud computing, big data management, internet of things, cyber security, enterprise computing & information systems management, and knowledge management.

**Gurjot Kaur** graduated in Electronics and Communication Engineering from Guru Nanak Dev University, Amritsar, India and completed her Masters in Electronics from UIET, Panjab University, Chandigarh, India. She is currently pursuing PhD in wireless communication from Dr. B.R. Ambedkar, NIT Jalandhar, India. Her research is focused in the areas of traffic management in real time networks, mobility in wireless networks, wireless ad hoc and sensor networks, cognitive networks and advances in mobile computing.

**Santosh Kumar** is an Assistant Professor in Computer Science and Engineering discipline at IIIT-NR. He received Ph.D. from IIT BHU Varanasi and M. Tech from BIT Mesra Ranchi. His research interest includes animal biometrics, computer vision, deep learning, pattern recognition, wireless sensors and Internet of Things (IoT). He has published over 30 journal (SCI-index journal) and five conference papers, one patent, and 34 book chapters in edited books (Springer publication and IGI publication). He is an active member of the IEEE, CSI, Computer Society and the Association for Computing Machinery. Recently, Dr. Santosh Kumar published the following books: 1. Animal Biometrics: Techniques and its Applications." Springer; 2. Multimodal Machine Learning: Algorithms and Applications (under review), Elsevier.

317

**Lalita Mishra** received a Bachelor of Technology degree in Computer Science & Engineering from Uttar Pradesh Technical University, Lucknow, India, in 2011 and Master of Technology in Computer Science & Engineering from International Institute of Information Technology, Bhubaneswar in 2015. Currently, she is working towards Doctor of Philosophy at Indian Institute of Information Technology, Allahabad in the Department of Information Technology.

**Neeraj Mogla** is currently working as Visual Insights & Analytics Manager at Nike Inc. global headquarters, USA. The author is MS in Computer Science from University of Akron, USA. The author has worked as Senior Application Engineer for UnitedHealth Group and Cognos Developer in R&D for eClinicalWorks LLC. The main area of expertise of author is data analytics, data sciences, and machine learning.

**Kavi Priya** is an Associate Professor in the Department of Information Technology at Mepco Sclenk Engineering College, Sivakasi, Tamilnadu, India. She obtained her Ph.D. from Anna University, Chennai, Tamilnadu, India in the area of routing in wireless sensor networks in the year 2018. She received her M.E. degree in Computer Science and Engineering from Anna University, Chennai, Tamilnadu, India in 2005, and her B.E. degree in Computer Science and Engineering from Madurai Kamaraj University, Madurai, Tamilnadu, India in 2002. Her research interests include wired and wireless networks. IoT, Data Mining, Big Data, Machine learning, soft computing etc. She has published many papers in national conferences, international conferences and refereed international journals.

**Virender Ranga** is an Assistant Professor in the Computer Engineering Department at the National Institute of Technology Kurukshetra, India. He received his PhD degree in 2016 from the National Institute of Technology Kurukshetra, Haryana, India. He has published more than 50 research papers in various international SCI Journals as well as reputed International Conferences in the area of Computer Communications. He has been conferred by Young Faculty Award in 2016 for his excellent contributions in the field of Computer Communications. He has been acted as a member of TPC in various International conferences of repute. He is a member of editorial board various reputed journals like Journal of Applied Computer Science & Artificial Intelligence, International Journal of Advances in Computer Science and Information Technology (IJACSIT), Circulation in Computer Science (CCS), International Journal of Bio Based and Modern Engineering (IJBBME) and International Journal of Wireless Networks and Broadband Technologies. Currently, he has been selected Guest Editor for a special issue to be published in the International Journal of Sensors, Wireless Communications and Control (Bentham Science Publication). He is an active reviewer of many reputed journals of IEEE, Springer, Elsevier, Talyor & Francis, Wiley, and InderScience. His current area of interest includes: wireless sensor and adhoc networks, Internet of Things, flying ad hoc networks, and software defined networking.

**Mamoon Rashid** is currently working as an Assistant Professor in the School of Computer Science & Engineering, Lovely Professional University, Jalandhar, India. The main area of interest is Cloud Computing, Big Data Analytics, Machine Learning and Neuroscience. The author has published several research papers in Cloud Computing, Big Data Analytics and Machine Learning indexed in Scopus and Web of Science based Journals.

**Urvashi Sangwan** has done B.Tech. in ECE from Guru Jambheshwar University of Science and Technology, Hisar, Haryana. She has completed her masters from Dr. B.R. Ambedkar, NIT Jalandhar. Her research interests are wireless sensors, Body Area Networks and wireless communication.

**Vignesh Saravanan** is a Assistant Professor in the Department of Computer Science and Engineering at Ramco Institute of Technology, Rajapalayam, Tamilnadu, India. He is pursing his Ph.D. from Anna University, Chennai, Tamilnadu, India in the area of Routing and Security in Wireless Sensor Networks from the year 2019. He received his M.Tech. degree in Information Technology from Mepco Schlenk Engineering College, Sivakasi, Tamilnadu, India in 2015, and his B.tech. degree in Computer Science and Engineering from Kalasalingam University, Tamilnadu, India in 2012. He has published one paper in international conference and one refereed international journal.

**Udai Shanker** is presently Professor in the Department of Computer Sc. & Engineering of M. M. M. University of Technology, Gorakhpur-273010.

**Diwankshi Sharma** received her B.Tech in Electronics and Telecommunication Engineering from Amritsar College of Engineering and Technology, Amritsar, Punjab, India in 2013 and is currently doing M.Tech at School of Electronics and Communication Engineering from Shri Mata Vaishno Devi University, Kakryal, Katra, (J. & K.). Her research interest includes Wireless Networking, Cryptography & Network Security. She has published one paper in International Journal of Simulation: Systems, Science and Technology.

**Pratik Shrivastava** is a research scholar at M.M.M.U.T., Gorakhpur - Department of Computer Science & Engineering. Her research interests include Replication Technique, Distributed Real Time Database System, Cryptography.

**Harjeet Singh** is presently working as an Associate Professor in the P.G. Department of Computer Science, Mata Gujri College, Fatehgarh Sahib, Punjab, India. The author has published many papers in indexed journals and currently guiding several research scholars at the doctorate level.

**Shreya Srivastav** is a student in National Institute of Technology, Hamirpur, Himachal Pradesh, India. She is currently pursuing her Dual Degree(M.Tech+B. Tech) in Computer Science and Engineering. She is doing her masters research in Cognitive Radio Networks. Her research interests include Cognitive Radio Networks, Wireless Sensor Networks, Machine Learning, Algorithms and Data Structures.

**Ashutosh Srivastava** received his B.Tech in Computer Science and Engineering from BIT, Merrut, Uttar Pradesh, India in 2006 and M.Tech & Ph.D. with specialization in Systems Engineering (Wireless Communication), Department of Electrical Engineering from Indian Institute of Technology (Banaras Hindu University) (IIT (BHU)), Varanasi, Uttar Pradesh, India in 2010 & 2016 respectively. Previously, he was the associate system engineer at IBM-SAP. Currently, he is working as Student Advisor at TBI-MCIIE IIT(BHU), Varanasi along with serving asa General Manager (honourable) at RST Ecoenergy Pvt Ltd. His research interest includes Adhoc Networking, Network Security, Software Defined Networking. He has published many papers and book chapters in the International/National journals and prestigious Conference Proceedings India as well as Abroad. He has served as session chair of the various workshop, seminar.

**Rochak Swami** is a PhD research scholar at the Department of Computer Engineering in National Institute of Technology, Kurukshetra, Haryana, India. She received her B.Tech degree (2012) in Computer Science and Engineering from Rajasthan Technical University, Kota, India and M.Tech degree (2015) in Computer Science from Birla Institute of Technology Mesra, Ranchi, Jharkhand, India. She is an ACM student member. Her current areas of interest include cyber security, security issues in Software defined networking, and DDoS attacks.

**Shirshu Varma** after completing a Ph.D. after served many reputed organizations like, BIT Mesra Ranchi, C-DAC Noida in the capacity of lecturer, Sr. lecturer & Principal project engineer. Presently working as a Professor in Indian Institute of Information Technology-Allahabad (IIIT-Allahabad). He has about 25 years of experience of teaching and research. He has published about 50 papers in international and national journals and conferences of repute and is the author of 04 book chapters. Number of citations of his papers is approximately 60. His area of work includes: wireless sensor networks coverage and connectivity, Sensor deployment and localization, Wireless sensor statistical routing, etc., WI-FI, WiMAX.

**Vijayalakshmi** is a Professor in the Department of Computer Science and Engineering at Ramco Institute of Technology, Rajapalayam, Tamilnadu, India. She obtained her Ph.D. from Anna University, Chennai, Tamilnadu, India in the area of network routing optimization using a genetic algorithm in the year 2008. She received her M.E. degree in Computer Science and Engineering from Madurai Kamaraj University, Madurai, Tamilnadu, India in 1999, and her B.E. degree in Computer Science and Engineering from Madurai Kamaraj University, Madurai, Tamilnadu, India in 1994. She guided 6 research scholars and currently guiding 3 research scholars in the area wired and wireless networks. IoT, data mining, Big Data, machine learning, soft computing, etc. She has published many papers in national conferences, international conferences and refereed international journals.

**Vikash** received the Bachelor of Technology degree in Computer Science & Engineering from Uttar Pradesh Technical University, Lucknow, India, in 2013, and Master of Technology degree in Computer Science & Engineering from Kamla Nehru Institute of Technology, Sultanpur, India, in 2015. Currently, he is working towards Doctor of Philosophy at Indian Institute of Information Technology, Allahabad in the Department of Information Technology.

# Index

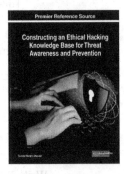

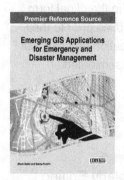